Innovations of Green Process Engineering for Sustainable Energy and Environment 2014

Topical Conference at the 2014 AIChE Annual Meeting

Atlanta, Georgia, USA
16-21 November 2014

ISBN: 978-1-5108-1251-2

Printed from e-media with permission by:

Curran Associates, Inc.
57 Morehouse Lane
Red Hook, NY 12571

Some format issues inherent in the e-media version may also appear in this print version.

Printed by Curran Associates, Inc. (2015)

AIChE
120 Wall Street, FL 23
New York, NY 10005-4020

Phone: (800) 242-4363
Fax: (203) 775-5177

www.aiche.org

Additional copies of this publication are available from:

Curran Associates, Inc.
57 Morehouse Lane
Red Hook, NY 12571 USA
Phone: 845-758-0400
Fax: 845-758-2634
Email: curran@proceedings.com
Web: www.proceedings.com

TABLE OF CONTENTS

Modeling and Experimental Validation in the Development of Chemical Hydrogen Storage Materials for Automotive Applications

Kriston Brooks, Rich Pires, Matt Westman
Kevin Simmons, Troy Semelsberger, Bart Van Hassel

2014 Annual AIChE Conference

Atlanta, GA
November 16, 2014

U.S. Department of Energy
Energy Efficiency and Renewable Energy
Bringing you a prosperous future where energy is clean, abundant, reliable, and affordable

Los Alamos NATIONAL LABORATORY

United Technologies Research Center

Pacific Northwest NATIONAL LABORATORY
Proudly Operated by Battelle *Since 1965*

Purpose of the Hydrogen Storage Engineering Center of Excellence

- Why Fuel Cell Cars?
  - Reduce US dependence on foreign oil
  - Reduce vehicle emissions
  - Increased efficiency
- Why not store neat hydrogen?
  - Gaseous--High pressure (200-700 bar)
  - Liquid --Low temperature (20 K)
  - Gravimetric and Volumetric Capacity
  - Dormancy
- What materials have been studied?
  - Adsorbents (MOF-5, AX-21)
  - Metal Hydrides (Mg_2Ni, TiCrMn)
  - **Chemical Hydrogen Storage**
- Center Efforts
  - Develop a system that meets the DOE Technical Targets

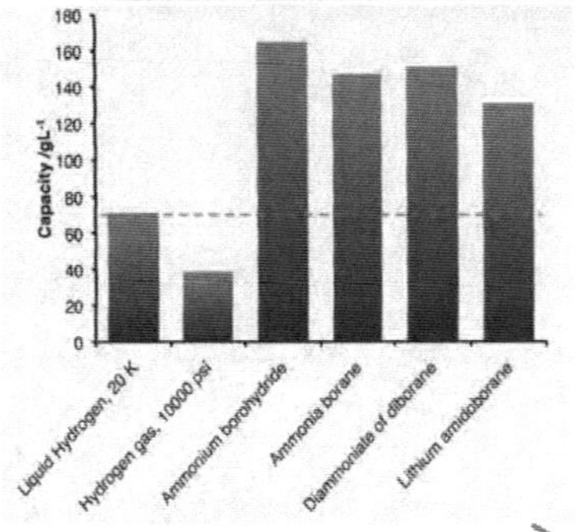

Pacific Northwest NATIONAL LABORATORY

2 Graph from http://energy.gov/eere/fuelcells/status-hydrogen-storage-technologies *Proudly Operated by* Battelle *Since 1965*

DOE Technical Targets

Storage Parameter	Units	2010	2017	Ultimate
System Gravimetric Capacity: Usable, specific-energy from H_2 (net useful energy/max system mass)	kWh/kg (kg H_2/kg system)	1.5 (0.045)	1.8 (0.055)	2.5 (0.075)
System Volumetric Capacity: Usable energy density from H_2 (net useful energy/max system volume)	kWh/L (kg H_2/L system)	0.9 (0.028)	1.3 (0.040)	2.3 (0.070)
Storage System Cost	$/kWh net ($/kg H_2)	TBD (TBD)	TBD (TBD)	TBD (TBD)
• Fuel cost	$/gge at pump	3-7	2-4	2-4
Durability/Operability:				
• Operating ambient temperature	°C	-30/50 (sun)	-40/60 (sun)	-40/60 (sun)
• Min/max delivery temperature	°C	-40/85	-40/85	-40/85
• Operational cycle life (1/4 tank to full)	Cycles	1000	1500	1500
• Min delivery pressure from storage system; FC= fuel cell, ICE= internal combustion engine	bar (abs)	5 FC/35 ICE	5 FC/35 ICE	3 FC/35 ICE
• Max delivery pressure from storage system	bar (abs)	12 FC/100 ICE	12 FC/100 ICE	12 FC/100 ICE
• Onboard Efficiency	%	90	90	90
• "Well" to Powerplant Efficiency	%	60	60	60
Charging / Discharging Rates:				
• System fill time (5 kg)	min (kg H_2/min)	4.2 (1.2)	3.3 (1.5)	2.5 (2.0)
• Minimum full flow rate	(g/s)/kW	0.02	0.02	0.02
• Start time to full flow (20°C)	s	5	5	5
• Start time to full flow (-20°C)	s	15	15	15
• Transient response 10%-90% and 90% - 0% [h]	s	0.75	0.75	0.75
Fuel Purity (H_2 from storage)	% H_2	SAE J2719 and ISO/PDTS 14687-2 (99.97% dry basis)		
Environmental Health & Safety:				
• Permeation & leakage	Scc/h	Meets or exceeds applicable standards		
• Toxicity	-			
• Safety	-			
• Loss of useable H_2	(g/h)kg H_2 stored	0.1	0.05	0.05

Useful constants: 0.2778 kWh/MJ; 33.3 kWh/kg H_2; 1 kg H_2 ≈ 1 gal gasoline equivalent.

st
ATORY
3 e 1965

HSECoE Approach

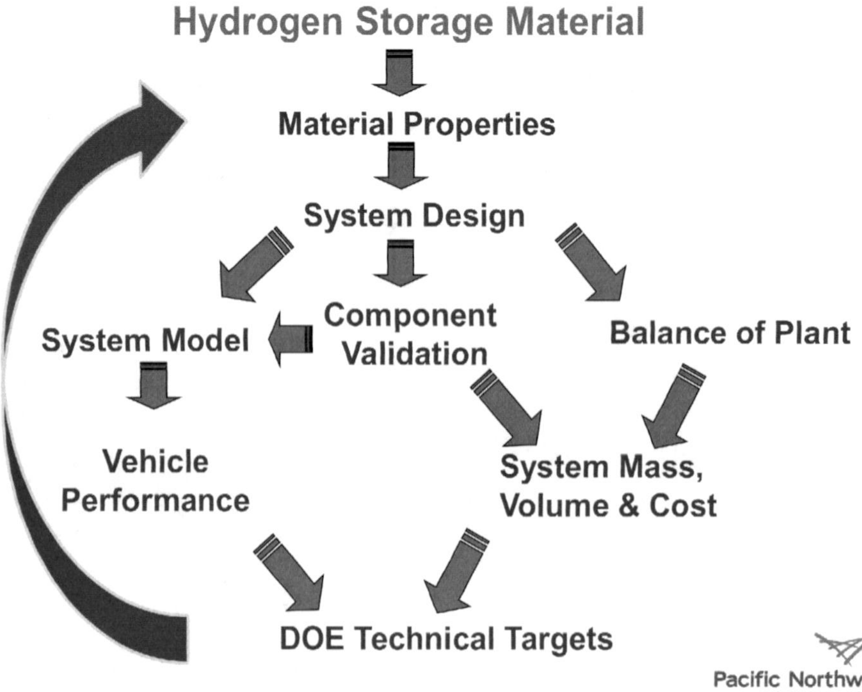

Pacific Northwest
NATIONAL LABORATORY

4 *Proudly Operated by* Battelle *Since 1965*

Chemical Hydrogen Storage Materials

- Characteristics
  - Hydrogen chemically bonded at densities greater than gaseous hydrogen
  - Off-Board regeneration
- Chemical Hydrogen Storage Materials of Interest
  - Ammonia Borane (NH_3BH_3)
    - 16 wt% H_2, ΔH_{rxn} = -22 kJ/mol H_2, ρ = 0.76 g/cc
    - White, waxy solid, stable in air at room temperature
  - Alane (AlH_3)
    - 10 wt% H_2, ΔH_{rxn} = +11.4 kJ/mol H_2, ρ = 1.49 g/cc
    - White, crystalline solid, readily decomposes in air or water
  - Slurries
    - Need for a flowable media
    - Silicone oil

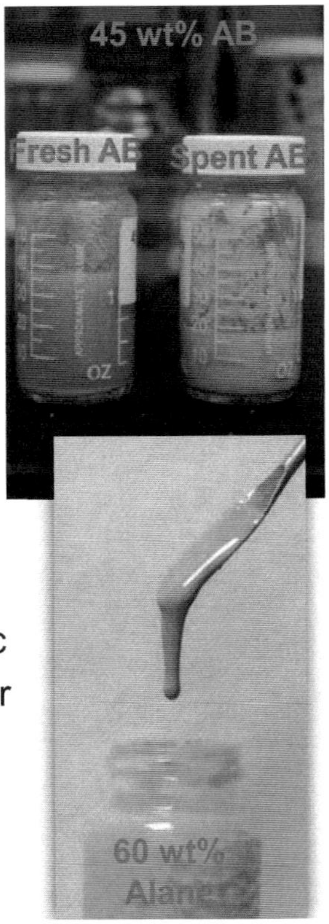

5

HSECoE Approach

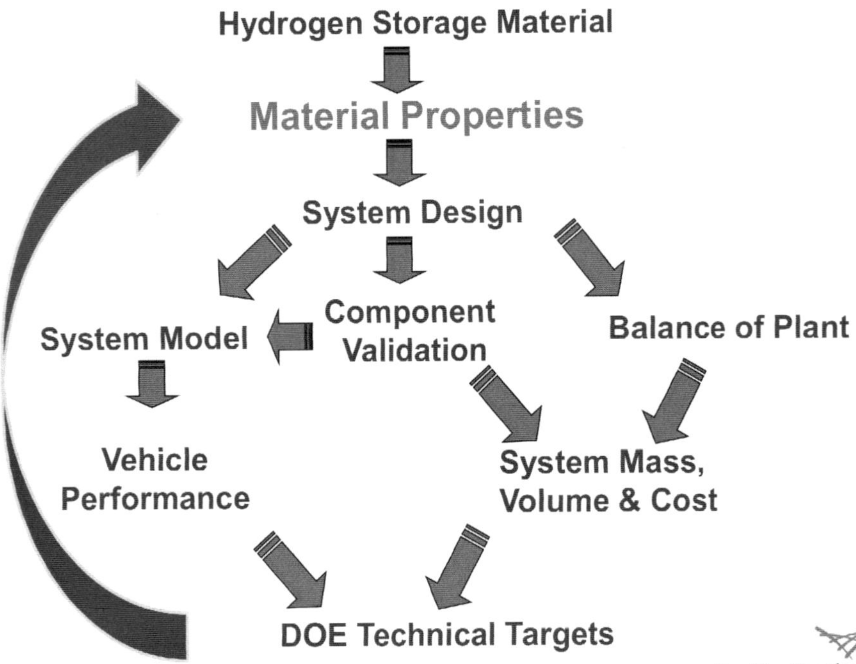

Pacific Northwest
NATIONAL LABORATORY
Proudly Operated by Battelle *Since 1965*

6

Material Properties

Reaction kinetics for slurry AB faster than solid AB

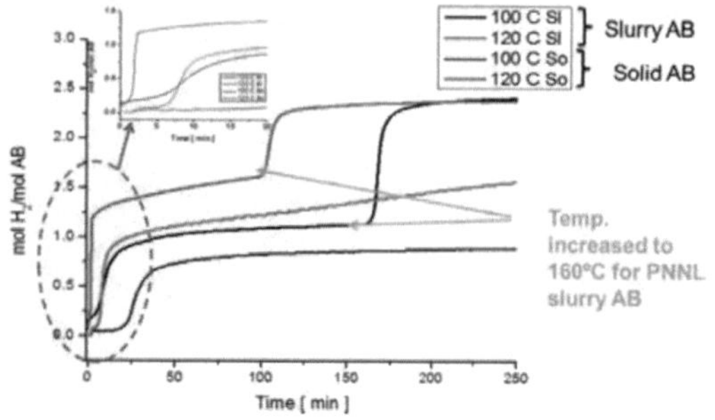

Viscosity linear at <35 wt%, rises exponentially at > 35 wt%

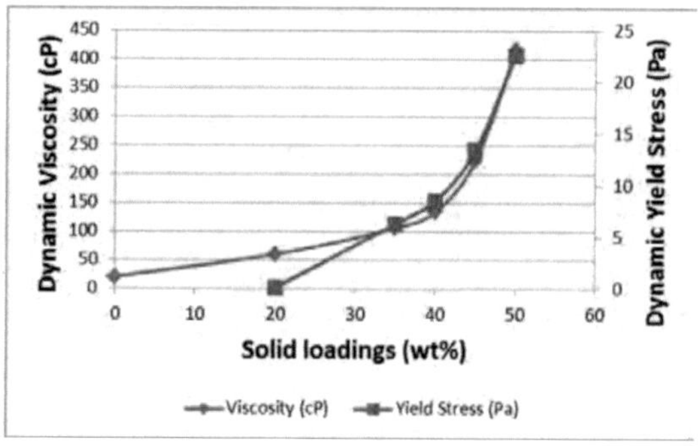

Weylchem AB with consistent distribution can produce higher solids-loaded slurries than Sigma-Aldrich

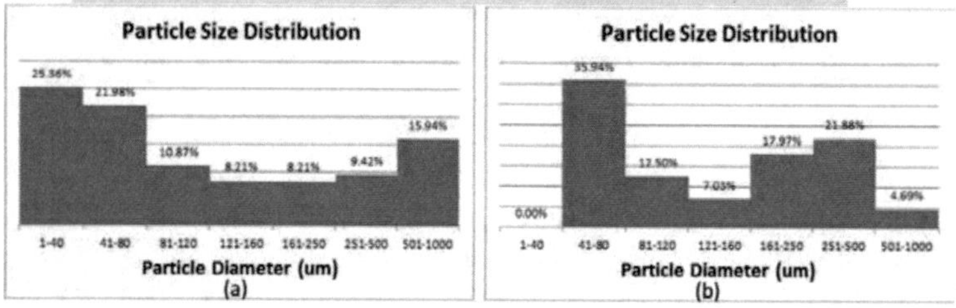

Pacific Northwest NATIONAL LABORATORY

Proudly Operated by Battelle *Since 1965*

HSECoE Approach

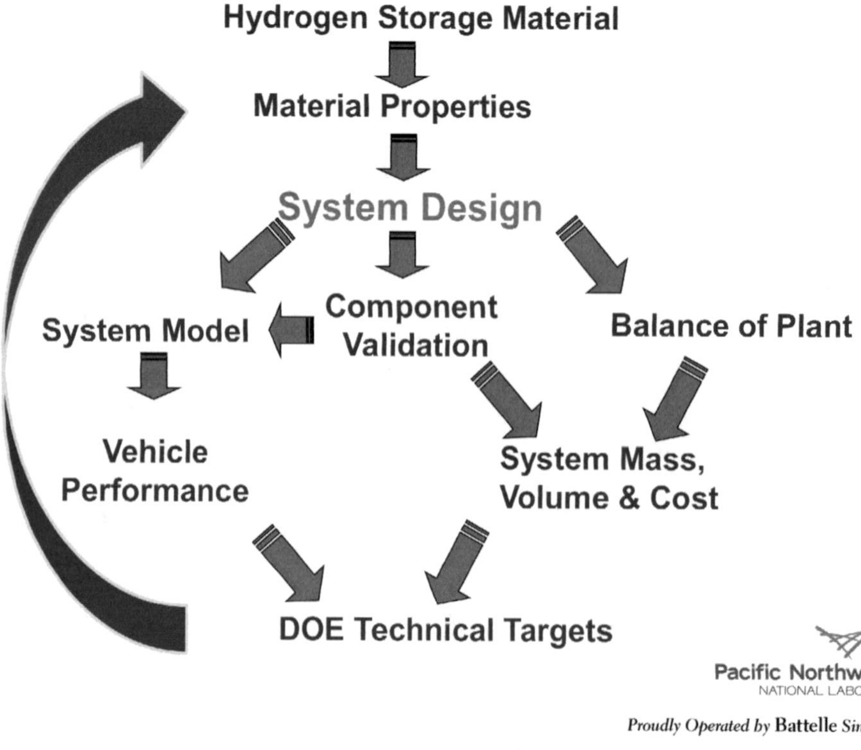

Pacific Northwest NATIONAL LABORATORY

Proudly Operated by Battelle *Since 1965*

System Design

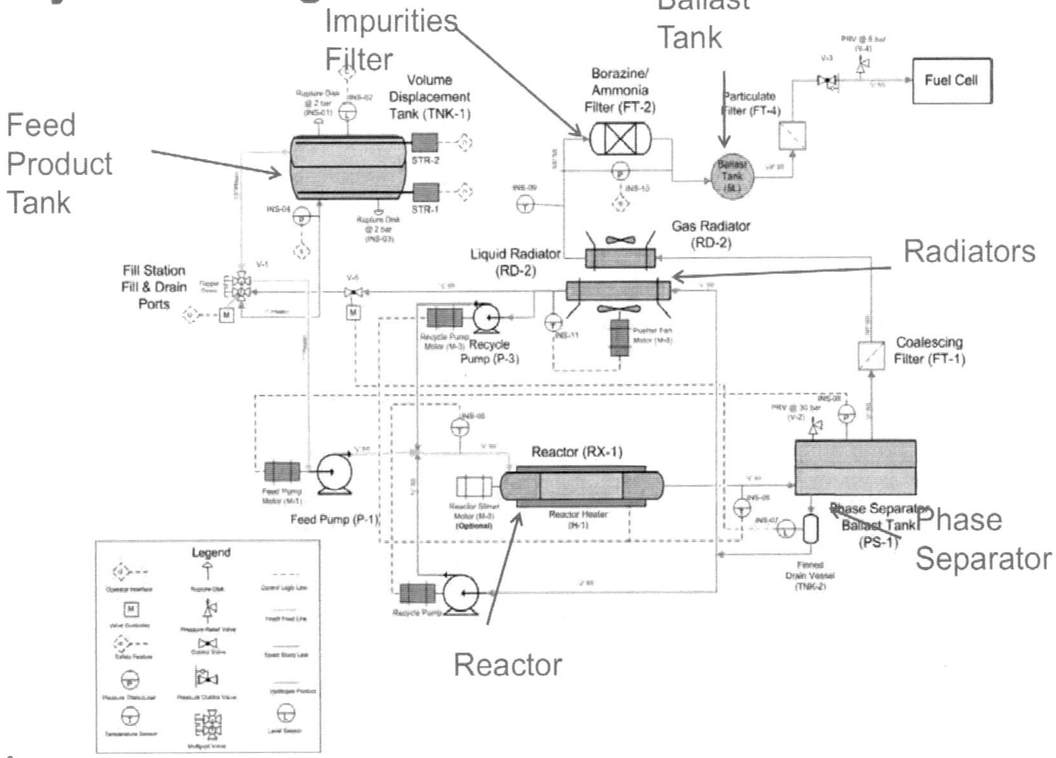

9

HSECoE Approach

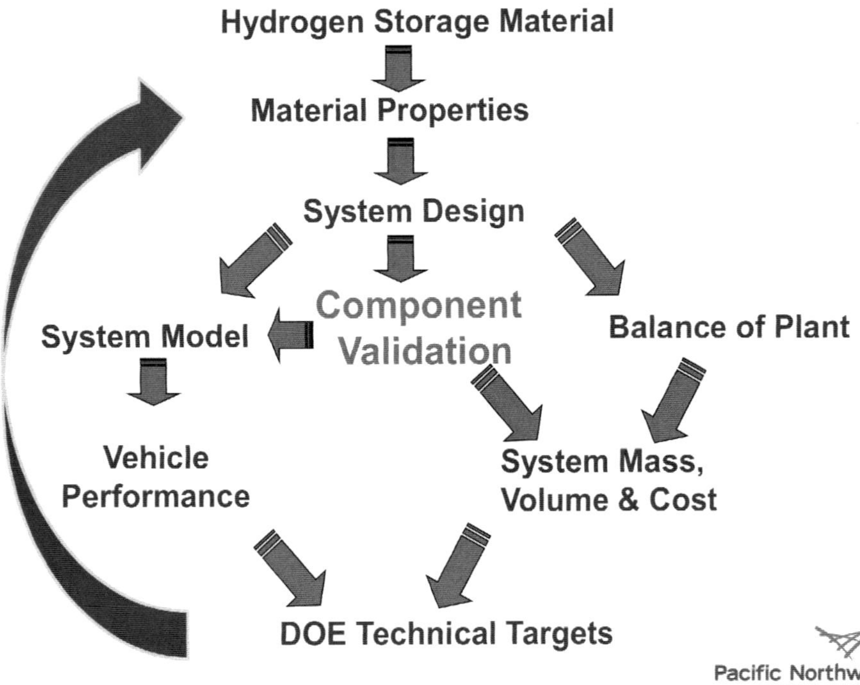

Component Validation Testing

= PNNL Performed Validation Testing

= LANL Performed Validation Testing

= UTRC Performed Validation Testing

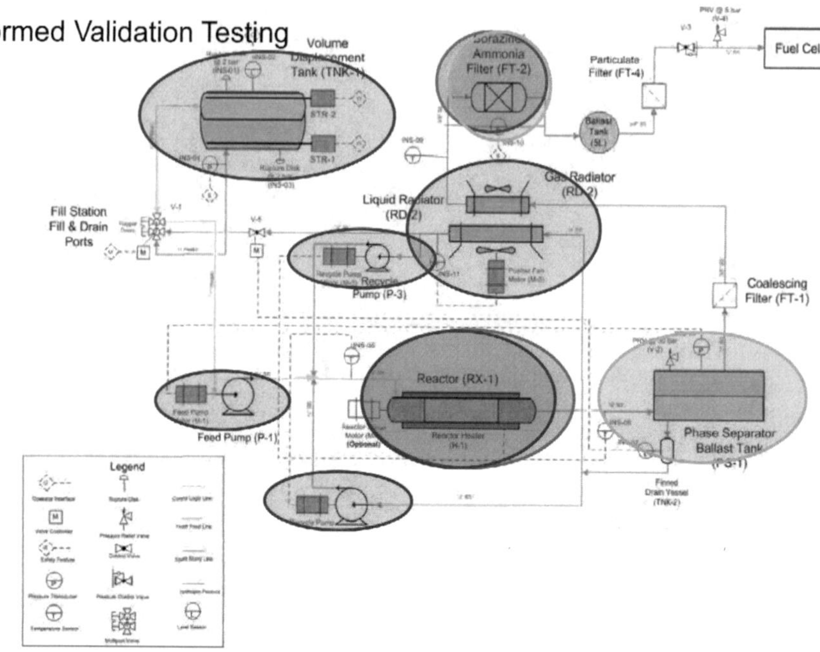

11

Component Validation

Flow Through Reactor Experiments

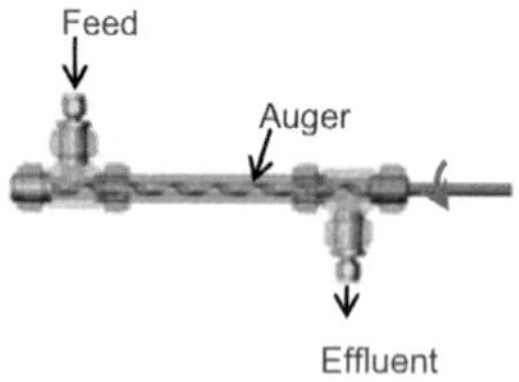

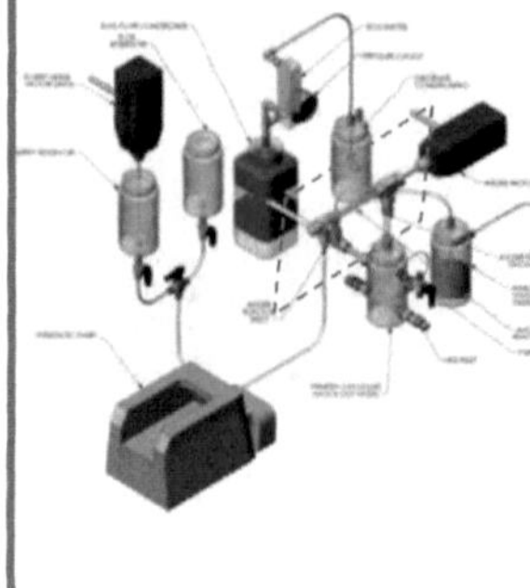

20-50 wt.% AlH_3 Successfully Slurries Tested

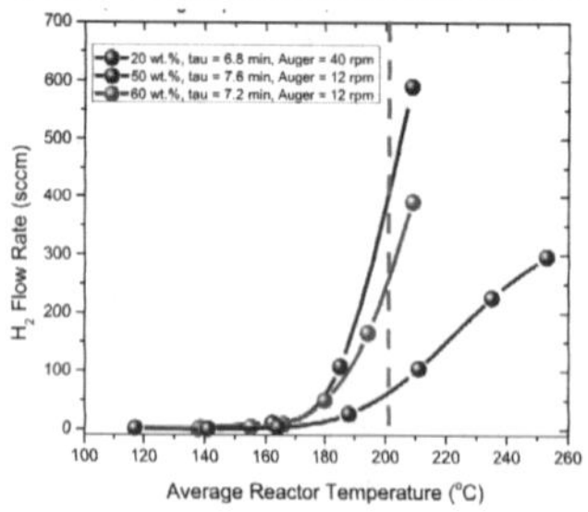

60 wt.% AlH_3 in MPF

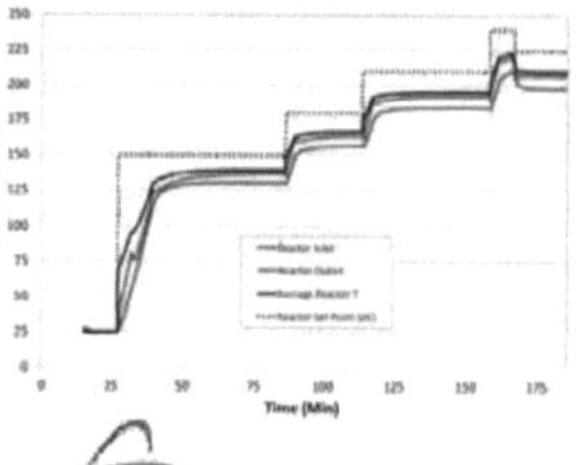

Reactor

Los Alamos NATIONAL LABORATORY

GLS Designed and Evaluated

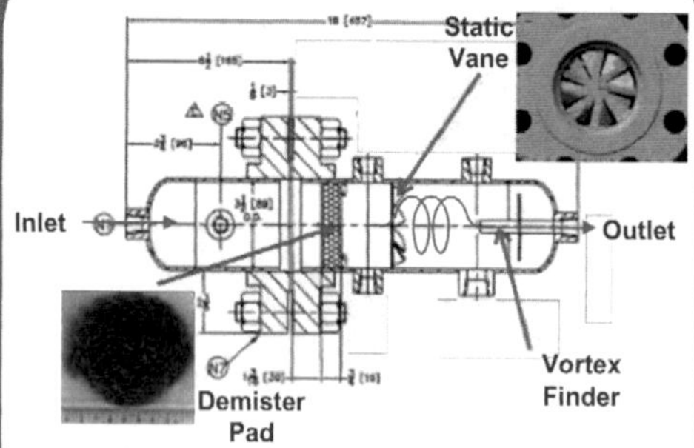

GLS shown to Effectively Eliminate oil Carryover Under Operational Conditions

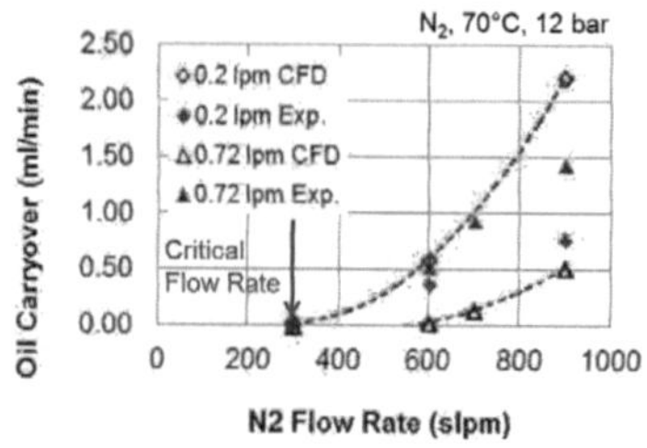

Phase Separator

United Technologies Research Center

12

Component Validation

Radiator

Radiator testing performed with full-scale prototype using polyimide simulant

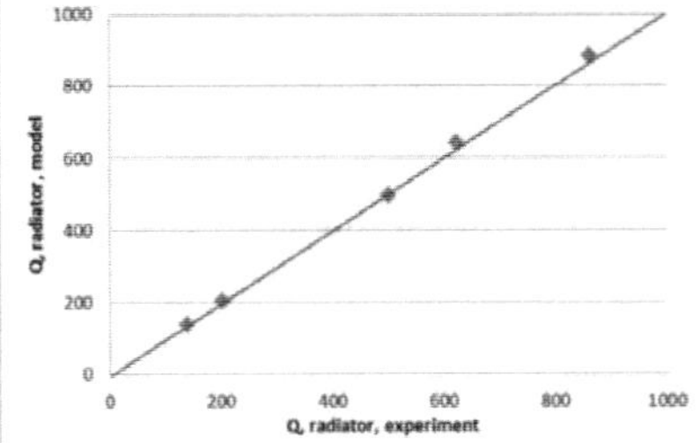

Experiment compared favorably to model

Slurry Pump

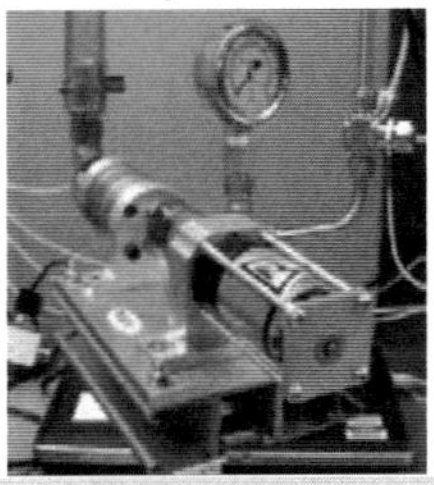

Pump worked for spent AB, but plugged with fresh AB

H_2 Clean-Up

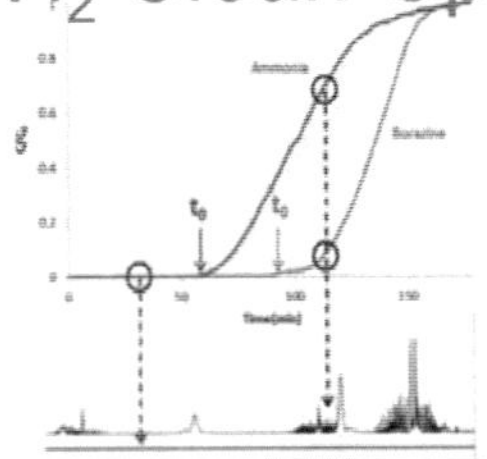

Ammonia and Borazine Filters Effectively Eliminate Gas Contamination

Tank

1/5th scale volume displacement tank validated membrane design, homogenization approach, and fill/drain requirement

HSECoE Approach

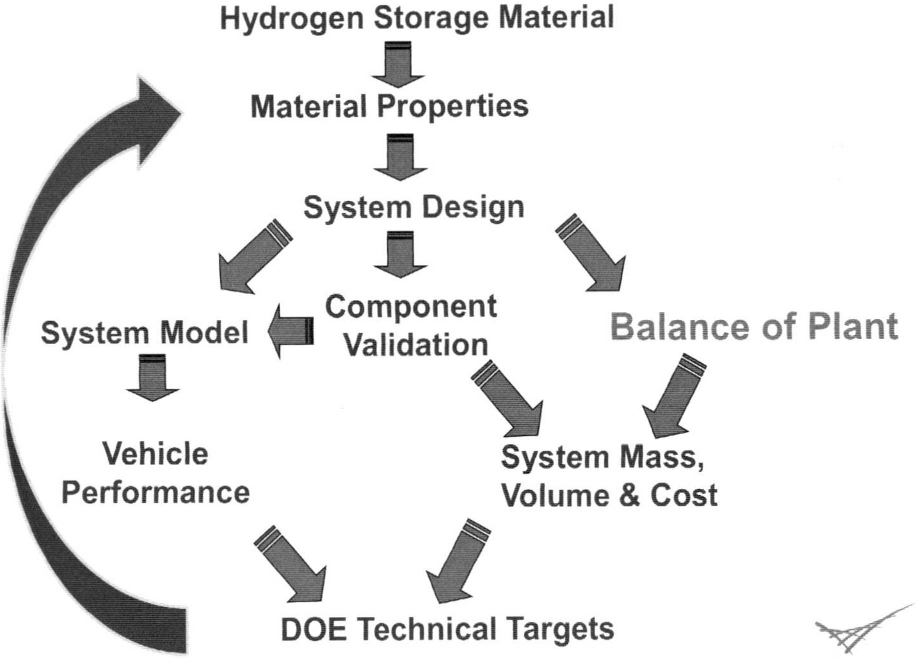

Pacific Northwest NATIONAL LABORATORY

Proudly Operated by Battelle *Since 1965*

Balance of Plant Component Selection

- Commercially available components when possible
- Minimize mass, volume and cost
- Designed for pressure and temperatures expected
- Designed for hydrogen and/or slurry compatibility

Item #	Description	Material	Wt (kg)	Vol (L)
Tanks and Tubing				
TNK-1	Volume Displacement Tank	High Density Polyethylene	6.2	86.0
NA	Fill and Drain Lines	10 ft of 1/2" Plastic	0.17	0.38
NA	Low T and P Lines	10 ft 3/8" Aluminum	0.12	0.2
NA	High T and P Lines	10 ft 3/8" Stainless Steel	0.38	0.22
INS-01	Rupture Disk		0.6	0.16
INS-02	Level Sensor for Volume Displacement Tank		0.6	0.16
INS-03	Rupture Disk		0.6	0.16
INS-04	Pressure sensor	316L SS	0.14	0.001
Feed Loop				
V-1	2 Multiport Valves with Actuator	Assured Automation	1.7	0.75
V-1	Flapper Valves		0.5	0.2
P-1	Feed Pump	KNF NF2.35	0.3	0.3
INS-05	Temperature sensor		0.1	0.02
RX-1	Reactor	SS tubing and stirrer	5	4
H-1	Reactor Heater		0.5	
INS-06	Temperature sensor		0.1	0.02
INS-07	Level Sensor for P/S		0.18	0.14
Recycle Loop				
P-2	Recycle Pump	KNF NF2.35	0.3	0.3

Item #	Description	Material	Wt (kg)	Vol (L)
Return Loop				
PS-1	Gas Liquid Separator	347/347L SS	3.2	3.7
INS-08	Pressure sensor	316L SS	0.14	0.001
V-2	Pressure Relief Valve		0.3	0.1
RD-2	Liquid Radiator	304 SS	2.08	2.9
RD-2	Liquid Radiator Header	304 SS	0.16	0.06
M-5	Liquid Radiator Fan Ultra Thin Line 12V Electric Fan (Puller)	Nylon	1	5.9
INS-11	Temperature sensor		0.1	0.02
V-5	Control Valve	Brass	1.7	0.75
Hydrogen Discharge				
FT-1	Coalescing Filter	SS	1.2	0.34
RD-2	Gas Radiator	304 SS	0.3	0.3
RD-2	Gas Radiator Header	304 SS	0.16	0.03
INS-09	Temperature sensor		0.1	0.02
INS-10	Pressure Switch		0.1	0.001
FT-2	H2 Clean-Up System		3.2	4
TNK-2	Additional Ballast Tank	Aluminum, L/D =4 , SF = 1.5	2.6	15
FT-4	Particulate Filter	SS	1.2	0.34
V-3	Pressure Regulator Gas		0.6	0.5
V-4	Pressure Relief Valve		0.6	0.16

15

Proudly Operated by Battelle *Since 1965*

HSECoE Approach

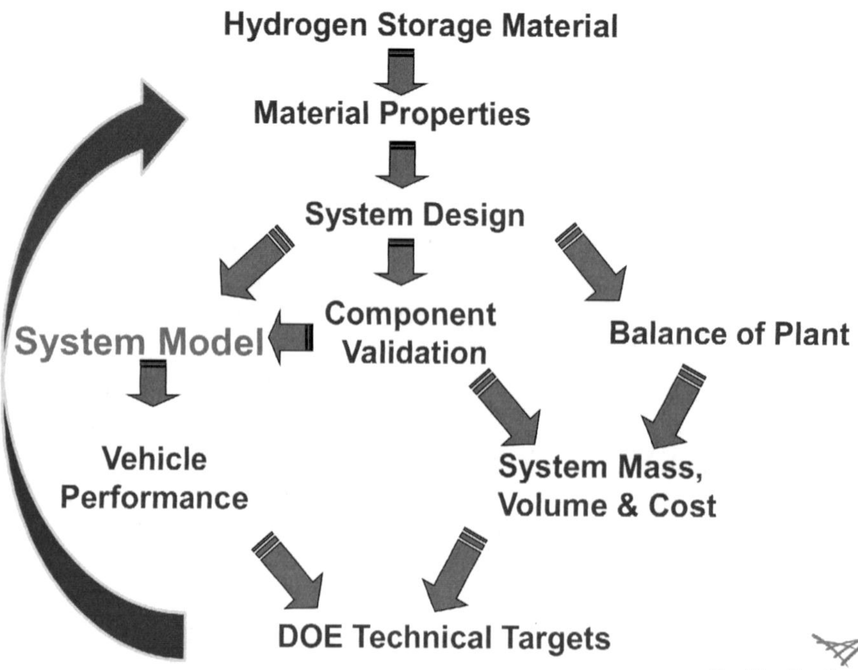

Pacific Northwest NATIONAL LABORATORY

16

Proudly Operated by Battelle *Since 1965*

System Modeling

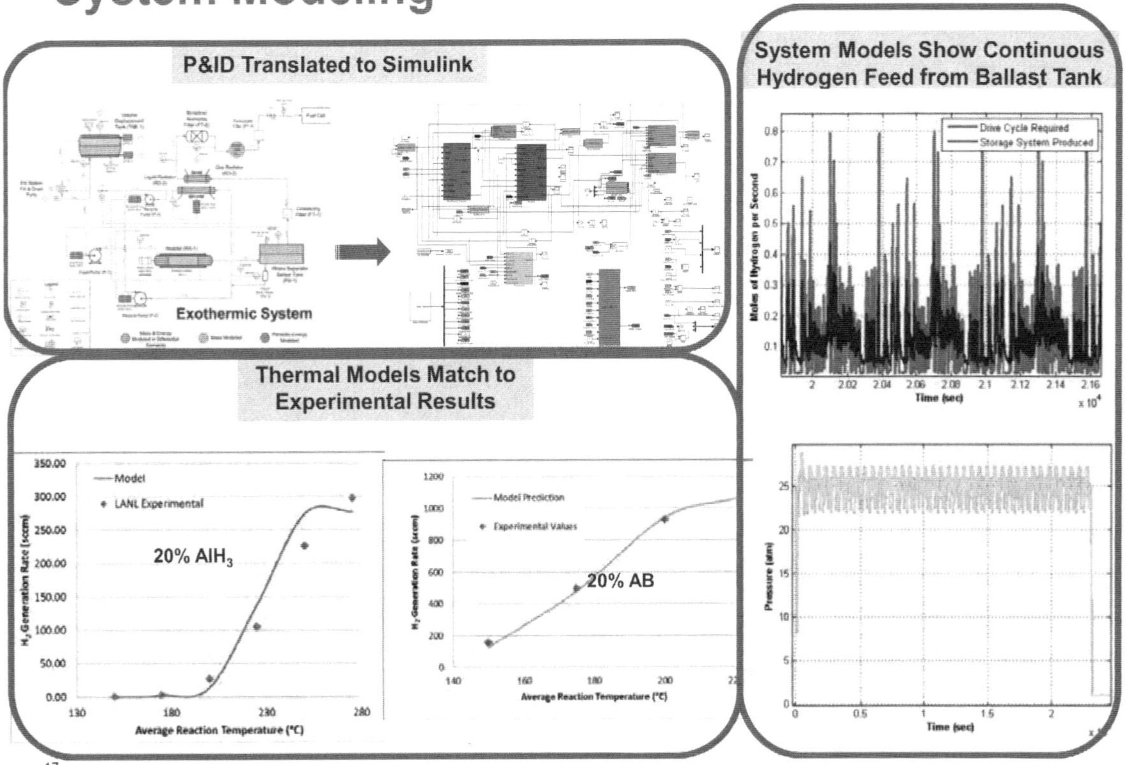

17

AB—US06 Drive Cycle Results

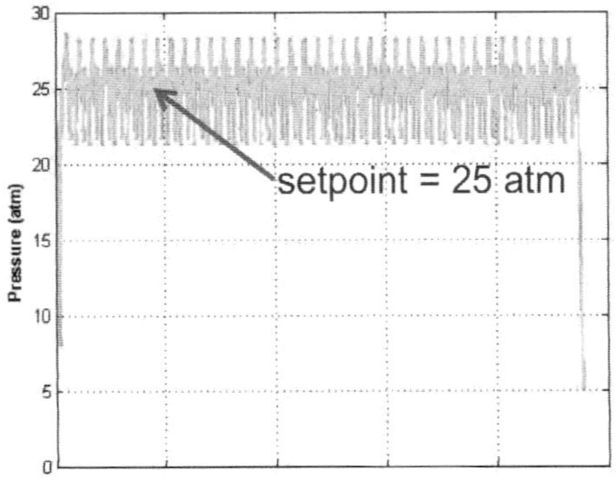

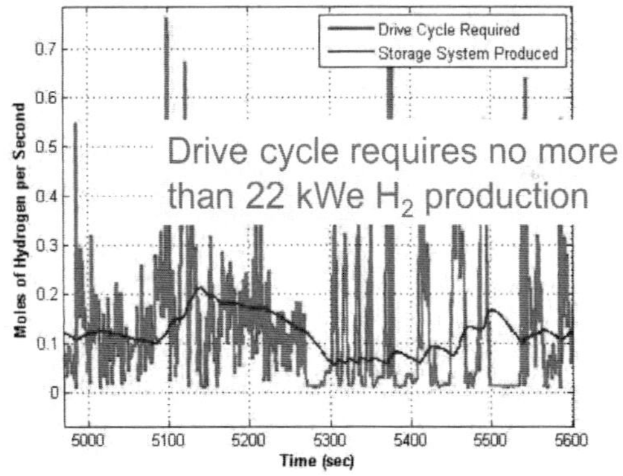

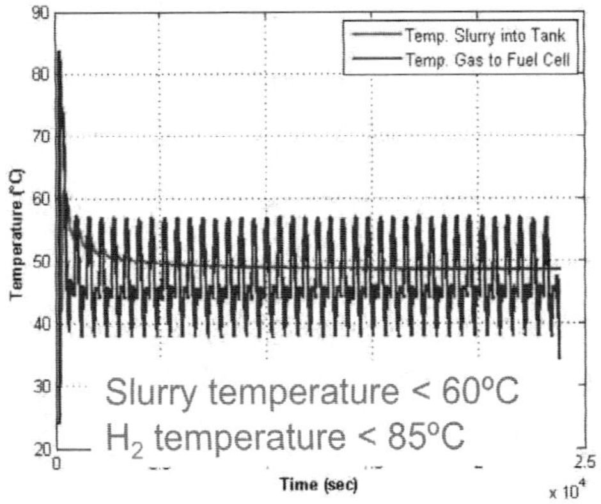

- 600 sec US06 drive cycle repeated 37 times or until stored 5.6 kg of H_2 exhausted
- Storage system must meet H_2 needs or drive cycle is stopped

Proudly Operated by Battelle *Since 1965*

Alane—UDDS Drive Cycle Results

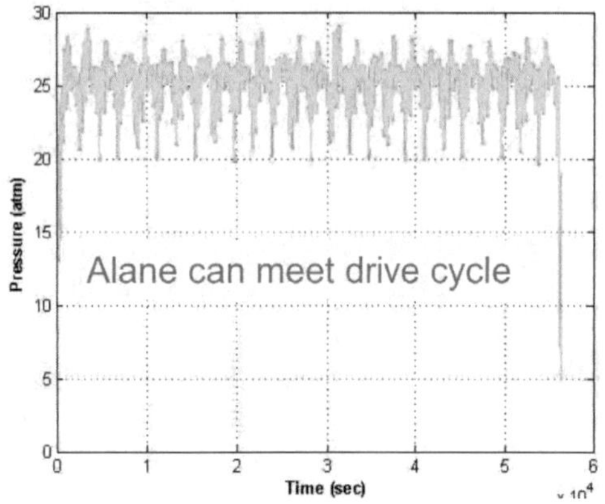

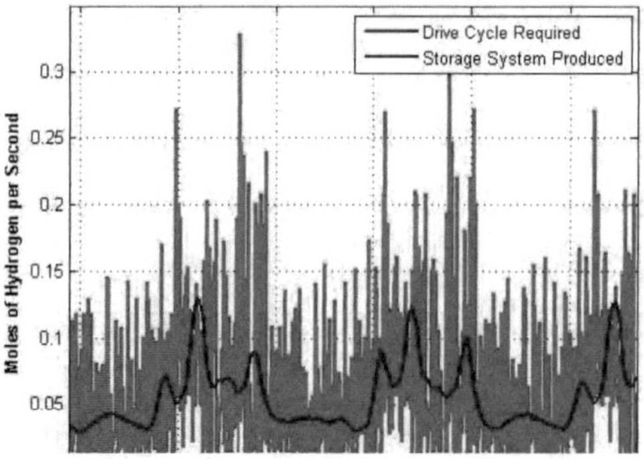

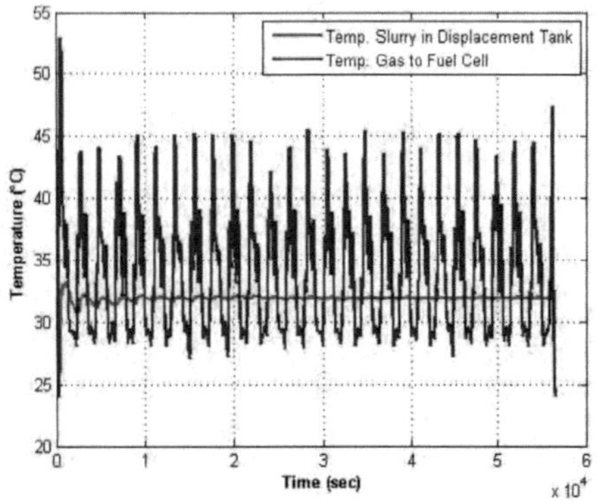

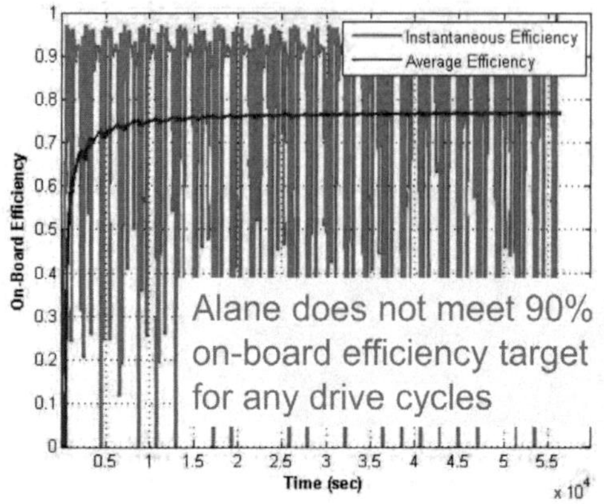

HSECoE Approach

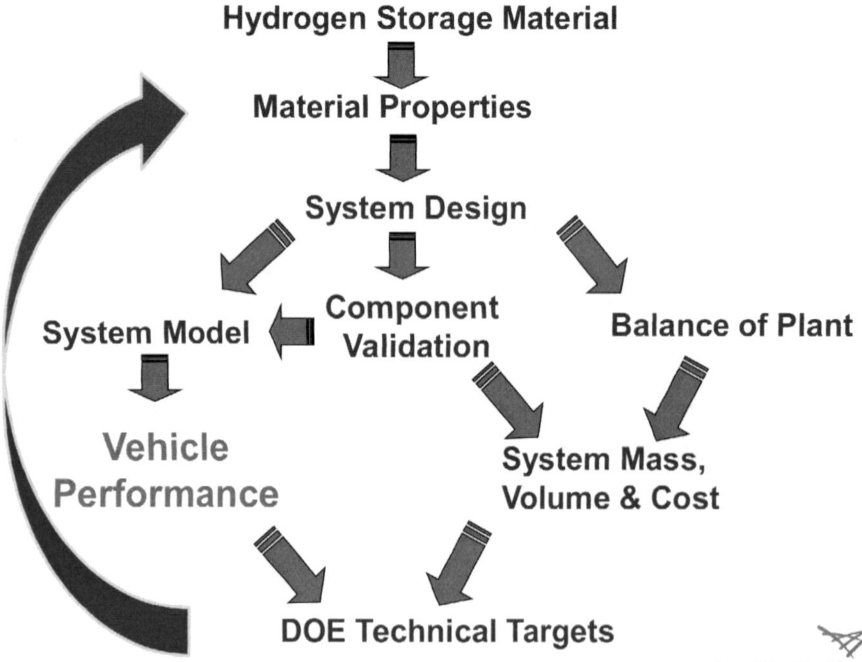

Vehicle Level Performance

www.hsecoe.org

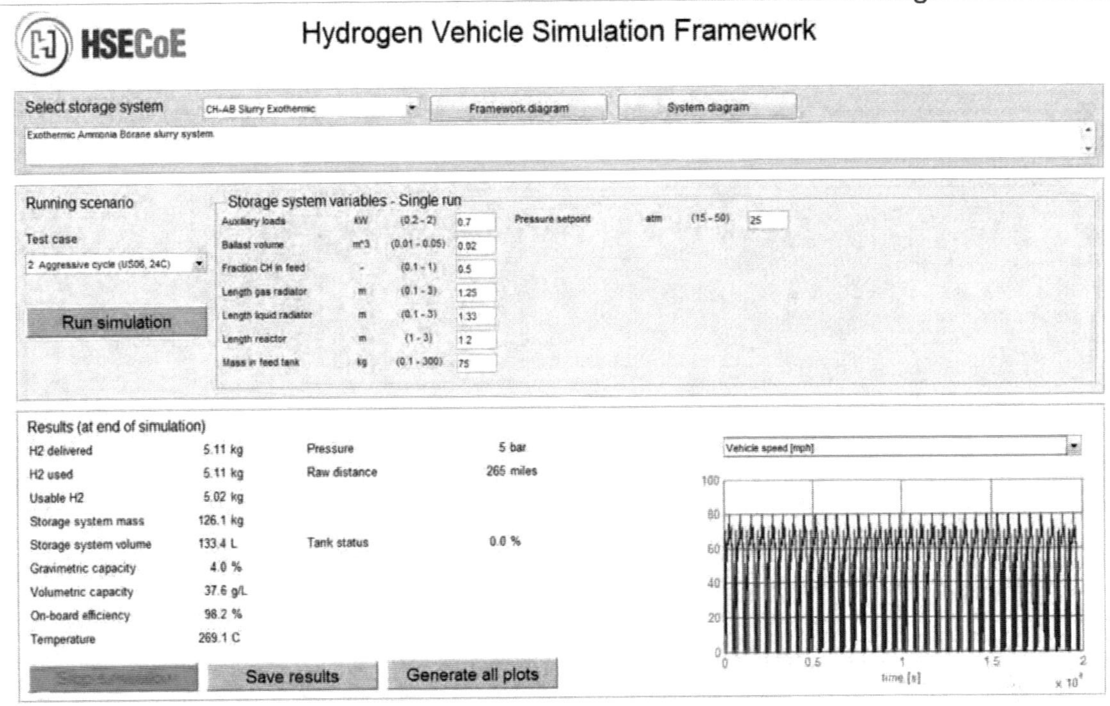

HSECoE Approach

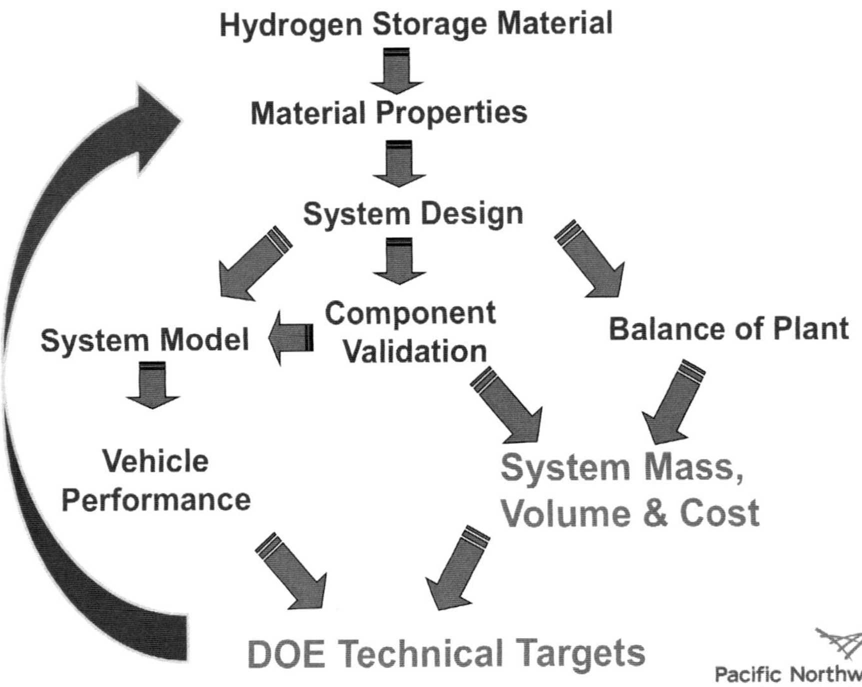

Pacific Northwest
NATIONAL LABORATORY

Proudly Operated by Battelle *Since 1965*

22

DOE Technical Targets

Ammonia Borane

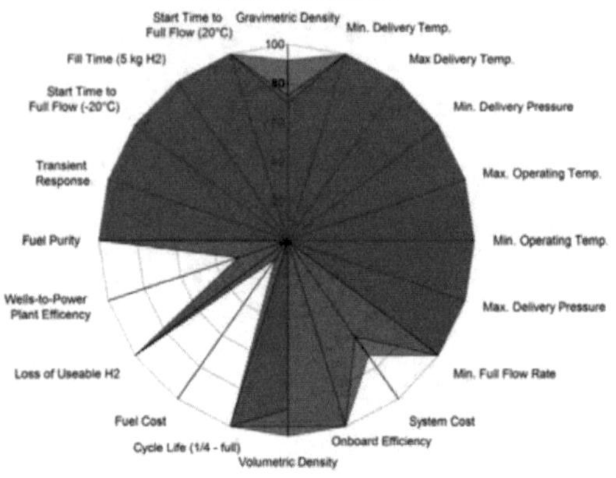

Alane

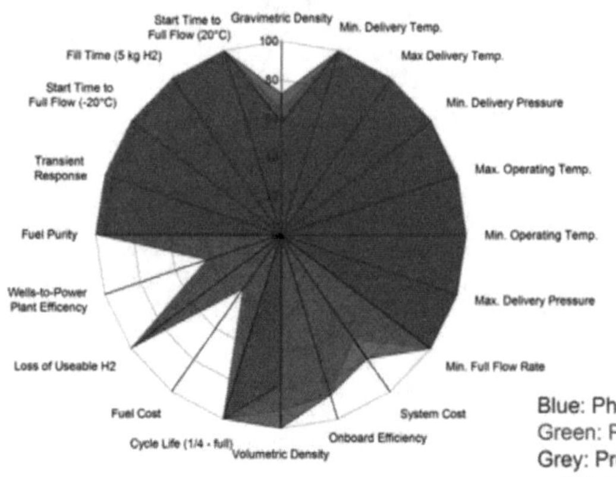

Blue: Phase 1
Green: Phase 2
Grey: Projected
(HSECoE Estimates)

50 wt.% slurries

AB Technical Hurdles: Fuel Cost, System Cost, WTPP, Gravimetric Density

Alane Technical Hurdles: Fuel Cost, System Cost, WTPP, On-board Efficiency, Gravimetric Density

Pacific Northwest
NATIONAL LABORATORY

23

Proudly Operated by Battelle *Since 1965*

HSECoE Approach

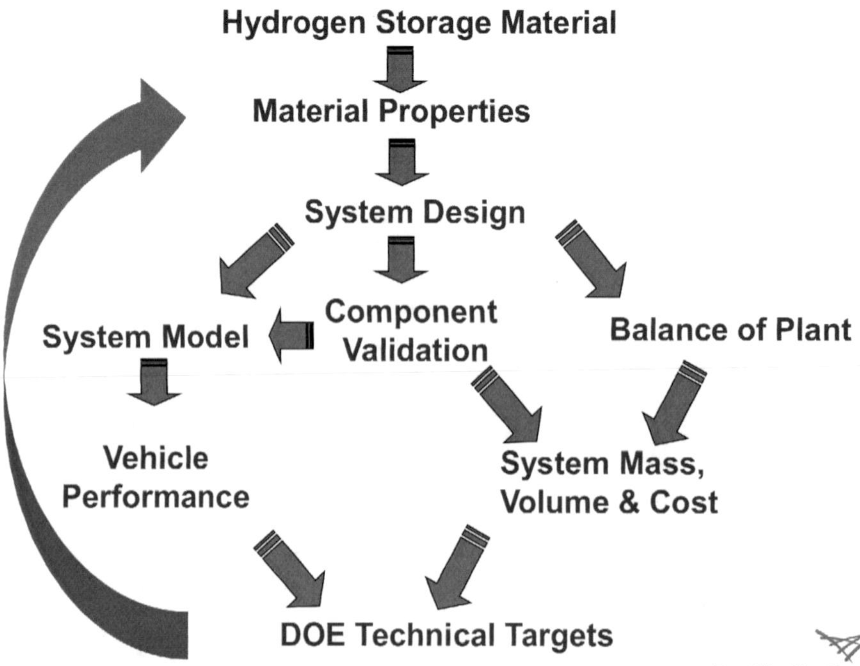

Pacific Northwest
NATIONAL LABORATORY

24

Proudly Operated by Battelle *Since 1965*

Model Sensitivity Analysis

- Use the model to evaluate characteristics other materials relative to the DOE Technical Targets
  - Based on current system design, slurry requires >11% hydrogen storage capacity (AB ~ 8%, alane ~ 6%) to meet gravimetric capacity
  - Heat of reaction < 10 kJ/mol H_2 to meet on-board efficiency

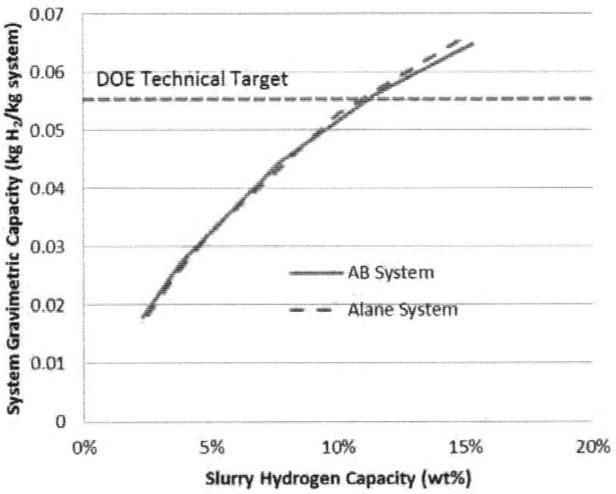

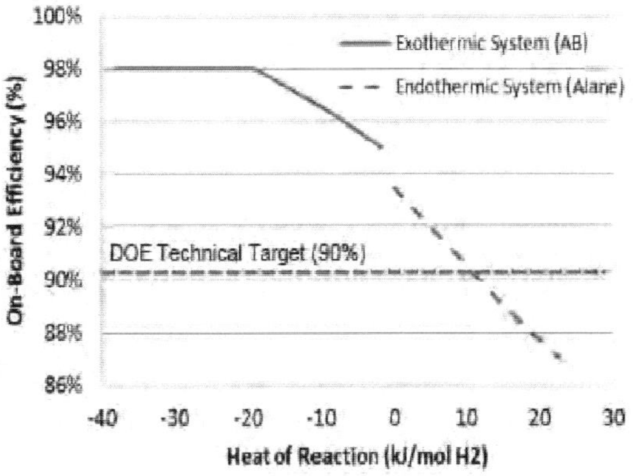

25

Develop Material Criteria Based on System Evaluation

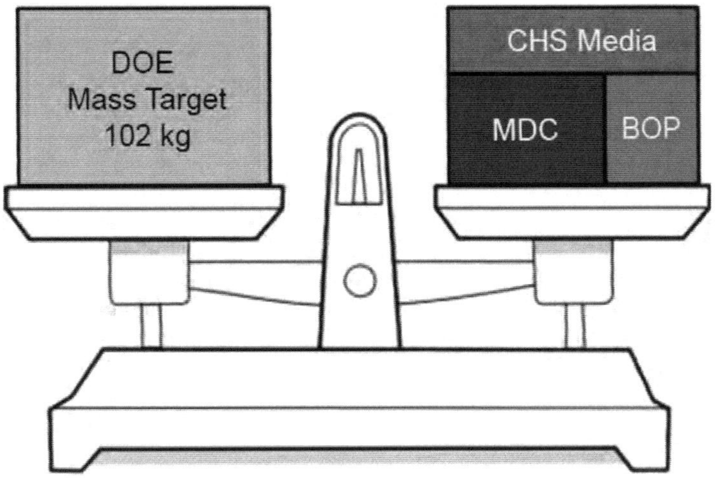

CHS Media = 102 - 14.5 - 21.8 = 65.7 kg

Material Parameters

Media Gravimetric Density

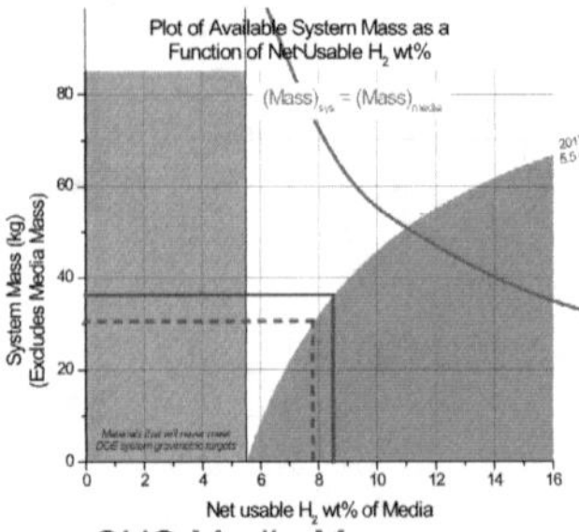

CHS Media Mass vs. System Mass

Material Density

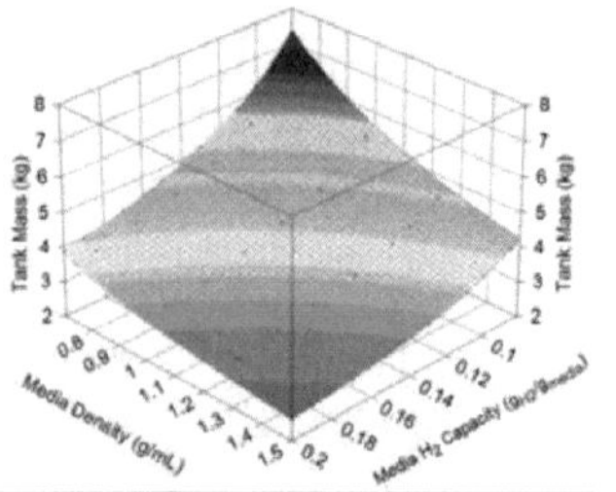

Size of Volume Displacement Tank

Slurry Kinetics

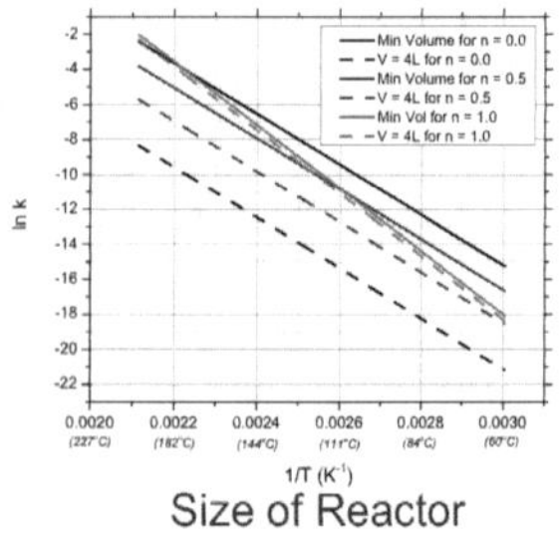

Size of Reactor Reaction at <60°C

Slurry Mass/Volume Fraction

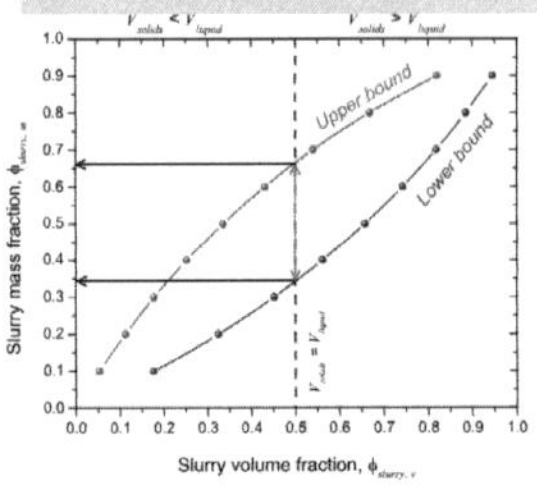

27

Flowable @ volume fraction < 0.5

Enthalpy of Reaction

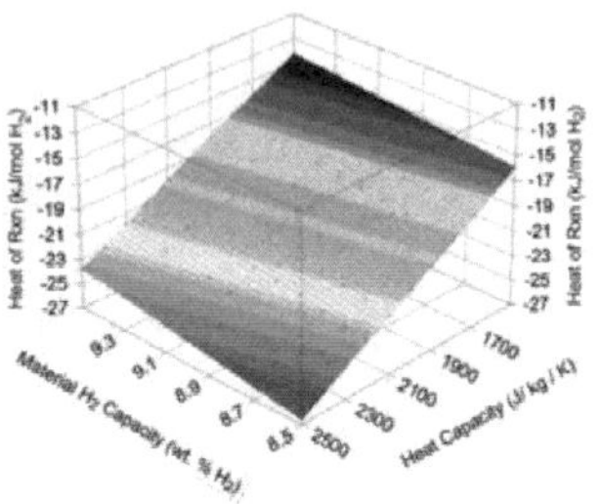

Size of Heat Exchanger

Impurities

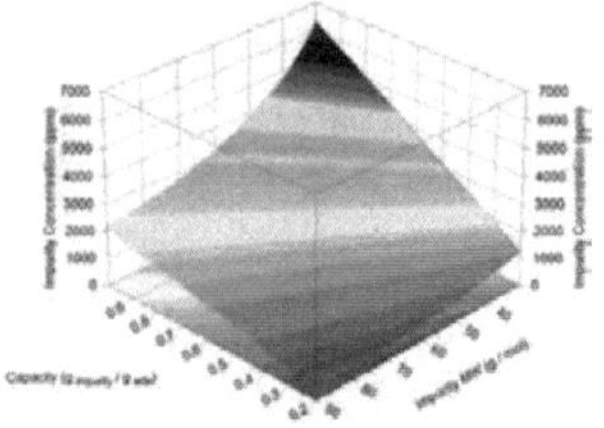

Size of Purification System

Acceptability Envelope for Chemical Hydrogen Storage Materials

Parameter	Symbol	Units	Range*	Influence	Assumptions
Minimum Material capacity (liquids)	γ_{mat}	$g_{H2} / g_{material}$	~ 0.078 (*0.085*)†	System	• System mass (excludes media) = 30.6 kg (36.3 kg) • 5.6 kg of H_2 stored • Liquid media (neat) • Media density = 1.0 g/mL
Minimum Material capacity (solutions)	γ_{mat}	$g_{H2} / g_{material}$	~ 0.098 (*0.106*)†	System	• System mass (excludes media) = 30.6 kg (36.3 kg) • Solute mass fraction = 0.35 ~ 0.80 • Solution density = 1.0 g/mL
Minimum Material capacity (slurries)	γ_{mat}	$g_{H2} / g_{material}$	~ 0.112 (*0.121*)†	System	• System mass (excludes media) = 30.6 kg (36.3 kg) • Non-settling homogeneous slurry • Slurry mass fraction = 0.35 ~ 0.70 • Slurry volume fraction = 0 ~ 0.5 • Slurry density = 1.0 g/mL
Kinetics: Activation Energy	E_a	kJ / mol	117-150	Reactor and Shelf life	• $V_{reactor} \leq 4$ L • Shelf life ≥ 60 days • Reaction order, n = 0 – 2
Kinetics: Preexponential Factor	A		$4 \times 10^9 - 1 \times 10^{16}$		
Endothermic Heat of Reaction	ΔH_{rxn}	kJ / mol H_2	≤ +17 (*15*)†	On-board efficiency	• On-board Efficiency = 90% • # Cold Startups = 4 • ΔT = 150 °C with no heat recovery • neat liquid (Cp = 1.6 J/g K) • Reactor mass = 2.5 kg SS (5.0 kg SS)
Exothermic Heat of Reaction	ΔH_{rxn}	kJ / mol H_2	≤ -27		• T_{max} = 250°C • Recycle ratio @ 50%
Maximum Reactor Outlet Temperature	T_{outlet}	°C	250	Heat Exchanger	• Liquid Radiator = 2.08 kg • Gas Radiator = 0.3 kg • Ballast Tank = 2.6 kg
Impurities Concentration	y_i	ppm	No *a priori* estimates can be quantified	Purification	• $m_{adsorbent} \leq 3.2$ kg
Media H_2 Density	$(\gamma_{mat})(\phi_m)(\rho_{mat})$	kg H_2 / L	≥ 0.07	Tank size System	• HD polyethylene tank ≤ 6.2 kg
Regen Efficiency	η_{regen}	%	≥ 66.6%	Well-to-Power Plant Efficiency	• On-board Efficiency = 90% • WTPP efficiency = 60%
Viscosity	η	cP	≤ 1500	Fill time Pump size On-board efficiency	None

Conclusions

- High H_2 content of Chemical Hydrogen Storage Materials makes them attractive
- The HSECoE developed an approach that helps materials developers determine if their chemical hydrogen storage materials will meet the DOE Technical Targets by
  - Developing a generic system design that includes component mass and volume (including BOP)
  - Establishing ranges of chemical parameters anticipated to meet targets based on the generic system design
  - Building a system model that can evaluate the performance of a chemical hydrogen materials in an automotive application
  - Validating the system design and model with experimental work
- Meeting all of the DOE Technical Targets will be extremely challenging

Pacific Northwest NATIONAL LABORATORY

29

Proudly Operated by Battelle *Since 1965*

Future Work

- Chemical Hydrogen Storage Material work for the HSECoE is coming to a close
  - Finalizing modeling effort—model will be published and available for download on the web
- Additional material development work to identify candidate materials that can meet the DOE technical targets
  - The modeling, validation tests, and development of acceptability envelope can help guide their efforts

Pacific Northwest NATIONAL LABORATORY

30

Proudly Operated by Battelle *Since 1965*

Acknowledgements

- Financial Support from DOE EERE Fuel Cell Technologies Office
  - DOE Management: Ned Stetson and Jesse Adams
- HSECoE Team Members

Questions?

Kriston P Brooks ☎: 509-372-4343 ✉: kriston.brooks@pnnl.gov

Pacific Northwest
NATIONAL LABORATORY
Proudly Operated by Battelle *Since 1965*

Separation processes in dehydrogenation of Liquid Organic Hydrogen Carriers

André Fikrt, Karsten Müller, Wolfgang Arlt

Friedrich-Alexander-University Erlangen-Nuremberg

Institute of Separation Science and Technology

Liquid Organic Hydrogen Carriers (LOHCs) are substances that are able to store hydrogen in a liquid, chemically bound form. As this reaction is completely reversible, the hydrogen can be regained at times of energy demand, while the carrier can be reloaded in times of excess energy. A LOHC systems not only consists of the chemical reactors, but a number of separation steps are necessary to fulfill all purity requirements. Starting from electrolysis the hydrogen has to be dried before hydrogenation. Afterwards the loaded carrier has to be degassed, as hydrogen is solved during the reaction. However, the most important separation task is the purification of the hydrogen produced in the dehydrogenation.

Although the dehydrogenation temperature is below the boiling point of the LOHC, the carrier is evaporating into the hydrogen gas phase. Taking dibenzyltoluene as a potential LOHC, nine moles of hydrogen gas are obtained from one mole of LOHC. Hence, despite the low vapor pressure crucial shares of the LOHC are evaporated. These impurities represent a loss of carrier and can harm subsequent reconversion systems, e.g. fuel cells.

Another important issue related to the evaporation is an increase of the heating demand of the already exothermic dehydrogenation due to the enthalpy of evaporation. This increased energy demand for heating the reaction can cause a significant decrease of the overall efficiency of the storage process.

To avoid this loss in efficiency, the separation system has to combine the cleaning of the gas with a heat recovery to lower the energy demand of the dehydrogenation. This is done by heat coupling of the condensation and the preheating of the hydrogen loaded LOHC.

Further condensation steps as well as subsequent fine cleaning can be necessary to reach the purity requirements. A prototype for a 12 kW hydrogen release unit was designed to demonstrate the feasibility of the concept.

Thermal properties and structural characterizations of new types of phase change material: Anhydrous and hydrated palmitic acid/camphene solid dispersions

Meng-Hsiu Chih, Yu-Hsiu Chiu, Yun Lee, Hung Lin-Lee and Tu Lee*

Department of Chemical and Materials Engineering, National Central University, 300 Jhong-Da Road, Jhong-Li City 320, Taiwan R.O.C.

ABSTRACT

There are two kinds of new phase change material (PCM) anhydrous and hydrous palmitic acid/camphene solid dispersions (PA1CA1). They are the mixing of 30.45 wt% amorphous crystalline camphene (CA), 57.32 wt% palmitic acid and the 12.23 wt% water, and they may have turned the matrix into more disorder because of their different bonding natures, molecular weights, and various molecular shapes and size. They could be proved by powder X-ray scattering and small-angle x-ray scattering. Their microstructures are found to be containing of nanometer-sized PA crystallites with lamellar periodicity, and their d spacing is 3.57 nm, dispersed in a partially amorphous plastic CA. The low-temperature differential scanning calorimetry and temperature-history method reveal that the PA phase apparently possesses a relatively high latent heat value inherited from the pristine crystalline PA of 229.7±0.1 kJ $kg^{-1}K^{-1}$. The relatively high specific heat of solid, C_{ps}, for anhydrous PA1CA1 of 2.17±0.06kJ $kg^{-1}K^{-1}$ is originated from the presence of disordered CA matrix. On the contrary, hydrated PA1CA1 could further increase the C_{ps} up to 2.61±0.01kJ $kg^{-1}K^{-1}$.

---

* Corresponding author. Telephone: +886-3-422-7151 ext. 34204. Fax: + 886-3-425-2296.
E-mail: tulee@cc.ncu.edu.tw

Fabrication of Photosystem I Protein Films with Improved Stability and Uniformity

Siyuan Yang, G. Kane Jennings, and David Cliffel
Department of Chemical and Biomolecular Engineering
Vanderbilt University, Nashville, TN

Photosystem I (PSI) is a nanoscale protein complex that can be incorporated into biohybrid devices to convert solar energy into usable electrical power. In prior research, our group has shown that PSI multilayer films that range in thickness from 0.3 - 3 μm generate photocurrent that increases with film thickness when deposited onto an electrode and exposed to a solution of redox molecules. However, the loss of PSI proteins exposed to the mediator solution may affect its performance in the long term. Also, the ability to prepare uniform PSI films in the thickness range of a few microns will facilitate integration with electrodes for solid-state biohybrid solar cells. Thus, the research in this presentation focuses on the improvement of stability and uniformity of PSI films. To improve the stability, different cross-linker molecules are introduced to conjugate proteins via covalent bonds to stabilize PSI multilayer films that were deposited onto electrodes. Results shows that 2-iminothiolane-cross-linked PSI films have the best performance in terms of both maintaining initial thickness and photocurrent after exposure to aqueous solutions for several days. To produce more uniform films, solid PSI is extracted from aqueous solution by centrifugation and then dissolved into volatile solvents with differing polarities. Slightly polar solvents provide better solvation of the PSI protein and straightforward deposition by both drop casting and spin coating. Importantly, PSI maintains its solar conversion function after exposure to these non-aqueous solvents. The work here provides guidelines for selecting an appropriate cross-linker for stabilizing PSI films as well as a good solvent for generating uniform PSI films.

390342 Enhancement of HCFC141b Hydration in Microemulsion with Nanoparticles As Additives for Cool Storage

Abstract:
Clathrate hydrate of refrigerant is regarded as a promising cool storage material. However, the low hydration rate and conversion of refrigerant impedes the application of the hydration technology. In this study, a kind of HCFC141b (CH3CCl2F)/Soya phosphatidylcholine (SPC)/Tween20/water microemulsion system was proposed to improve the solubility of HCFC141b in water and to promote the interfacial hydration reaction of refrigerant. HCFC141b/water microemulsion was prepared with as much as 26.4.1:100 mass ratio of HCFC141b to H2O. It was first observed that a large number of gas hydrate crystals formed rapidly and uniformly in the bulk solution under static condition. The amount of hydrates formed in the microemulsion was related to the amount of HCFC141b dissolved in water. The dissolution of HCFC141b in nanoscale microemulsion induced rapid and uniform hydrate formation and growth throughout the liquid. The induction and nucleation period of hydration in microemulsion was much shorter than that in HCFC141b/water crude emulsion and coarse mixture during several successive hydration processes. The memory effect induced a dramatic reduction in induction time and supercooling degree of the posterior hydration. The induction time of hydration was reduced by 52% and 36% compared to those in HCFC141b/H2O coarse mixture and crude emulsion systems, respectively. Furthermore, Al and Cu nanoparticles were incorporated with the microemulsion and the hydration was further promoted. The addition of the nanoparticals reduced substantially not only the induction time but also the supercooling degree of hydration. The addition of 0.1% (w/w) Al nanoparticles in the microemulsion reduced the induction time and supercooling degree of hydration by 62% and 54%, respectively. As a conclusion, the combination of suitable type and amount of nanoparticles in refrigerant/water microemulsion is an efficient way to promote refrigerant hydration.

Micro-Fuel Cell Combined Heat and Power Commercial Demonstration

Kriston Brooks, Atefe Makhmalbaf

2014 Annual AIChE Meeting

Atlanta, GA
November 16, 2014

U.S. Department of Energy
Energy Efficiency and Renewable Energy
Bringing you a prosperous future where energy is clean, abundant, reliable, and affordable

ClearEdge | POWER®
Power Your Independence.™

Pacific Northwest
NATIONAL LABORATORY
Proudly Operated by Battelle *Since 1965*

Purpose of the DOE's Market Transformation in the Fuel Cells Technology Office

- What it does . . .
  - Identify near term niche markets where fuel cells have potential
  - Work with stakeholders to develop activities in those areas
  - Collect data in real-world applications and provide unbiased analysis
  - Evaluate the business case for niche markets based on demonstration results
  - Present the results to the community
- What it hopes in return . . .
  - Support commercialization
  - Increase fuel cell production volume
  - Overcome non-technical roadblocks to commercialization

Pacific Northwest
NATIONAL LABORATORY
Proudly Operated by Battelle *Since 1965*

2

Comparison to Other Studies

- Scale
  - Micro-CHP is a unique range: 5-50 kWe
  - Other FCS/CHP manufactures focus on:
    - Large-scale industrial / commercial applications: >100 kWe
    - Residential Market: < 7 kWe
- Application
  - Small commercial buildings
- Duration
  - 3 year evaluation period as compared to 3-6 months typically done previously
  - Allows us to track system degradation and system development

Systems Studied

ClearEdge Power

- 5 kWe fuel cell
- 5.5 kWt hot water at 40-50° C
- Hydrogen from reformed natural gas

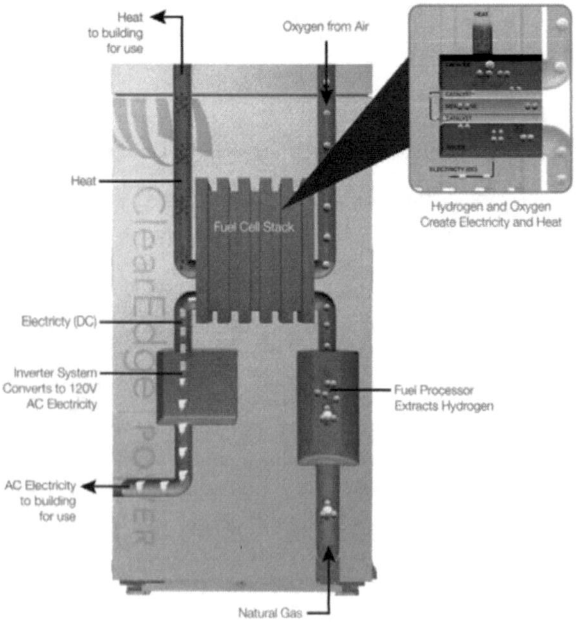

Phase I and Phase II Technology

	CE5	M5
Operation Period	Sept. 2011-Aug 2013	July 2013-July 2014
Stack Technology	High Temp PBI	Phosphoric Acid
Anticipated Lifetime	5 years	10 years
Footprint	1X	2X
Maintenance Access	All sides	Front access only
Cooling	Air Cooling—Outdoor	Glycol Cooling—Indoor
Power	Standby upon power loss	Grid independent, load following
Upgrades	8 units in mid-2012	All Units

Micro-CHP Fuel Cell System Case Study

Partner/Site	Location	Number of FCSs
College	Portland, OR	2
Nursery	Corona Del Mar, CA	3
Recreation	Oakland, CA	5
Grocery	San Francisco, CA	5
Total		15

Operational Targets

Average Net Electric Power Output [kWe]	Average Net Heat Recovery [kWt]	Temperature to Site [°C]	Average Net System Electric Efficiency[(a)] [%]	Average Net Heat Recovery Efficiency[(a)] [%]	Overall Net System Efficiency[(a)] [%]
5.0	**5.5**	**up to 65**	**36**	**40**	**76**

(a) Efficiencies are based on higher heating value (HHV).

Project Timeline

Events/Changes	Q4	FY12 Q1	FY12 Q2	FY12 Q3	FY12 Q4	FY13 Q1	FY13 Q2	FY13 Q3	FY13 Q4	FY14 Q1	FY14 Q2	FY14 Q3	FY14 Q4
Initial Deployment (CE5)	10	5											
Set-Point Change (5 kWe--> 4 kWe)		15											
BOP Upgrades (CE5)				8									
Units Shut down for major facility renovation				2									
Units Shutdowns for New M5 Upgrade									7	6			
New System Upgrade (M5)									2	1	7		
End of Data Analysis												10	0

(Column heading: Number of Units Involved in Each Event)

Number of Units Collecting Data

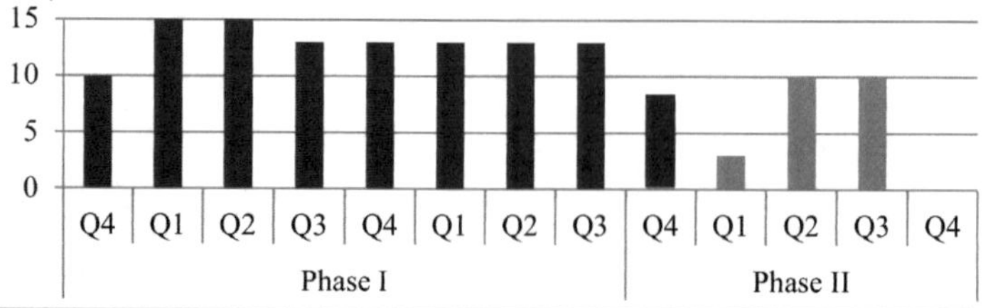

■ CE5 Units Operating

■ M5 Units Operating

Pacific Northwest NATIONAL LABORATORY

Proudly Operated by Battelle *Since 1965*

7

Engineering Data Analysis

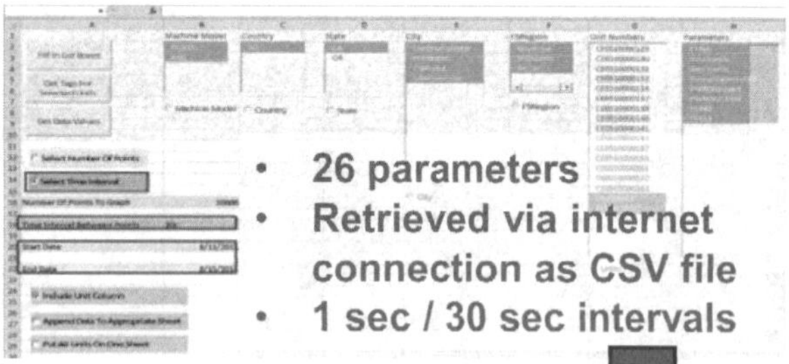

- 26 parameters
- Retrieved via internet connection as CSV file
- 1 sec / 30 sec intervals

- Natural Gas Usage (slpm)
- Electricity Produced (kWe)
- Estimated Heat Produced (kWt)
- Water Temperature to Site (°C)

R Statistical Computing Software

$$Net\ Heat\ Recovery\ Efficiency\ [\%] = \frac{HeatGenerated}{\dot{m}_{NG} * \Delta H_{NG}}$$

$$Net\ Electric\ Efficiency\ [\%] = \frac{Net\ Electric\ Output}{\dot{m}_{NG} * \Delta H_{NG}}$$

- Electrical and thermal efficiency (HHV)
- Overall efficiency (HHV)
- Overall Availability (> 1 kWe)
- Heat/Electricity utilized by the facility (2 sites)

Pacific Northwest NATIONAL LABORATORY

Proudly Operated by Battelle *Since 1965*

8

Failure Analysis

Date	General Cause of Shutdown	CSA #	Runtime To Shutdown	Time To Restart
12/13/2011	Site/Facilities	1	NA	
2/12/2012	CE5 Part Premature Failure	1	1436	
2/22/2012	CE5 Part Premature Failure	2	206	
4/20/2012	CE5 Part Premature Failure	2	1362	
5/23/2012	CE5 Part Premature Failure	[illegible]	699	
8/30/2012	CE5 Part Premature Failure	3	2370	
8/31/2012	CE5 Part Premature Failure	4	2378	
10/23/2012	CE5 Part Premature Failure	5	1152	54
11/26/2012	CE5 Part Premature Failure	6	771	34
2/4/2013	Site/Facilities	6	1664	46
5/16/2013	CE5 Part Premature Failure	6	2382	9
7/15/2013	CE5 Part Premature Failure	7	1415	32
9/5/2013	CE5 Part Premature Failure	8	1235	21
9/9/2013	CE5 Part Premature Failure	8	76	26

Quarterly Summary:
Maintenance Events
General Shutdowns

Spreadsheet Analysis

System Downtime

Premature part failure, human error, preventive maintenance, facility downtime

Pacific Northwest NATIONAL LABORATORY

Proudly Operated by Battelle *Since 1965*

Phase I Time Periods

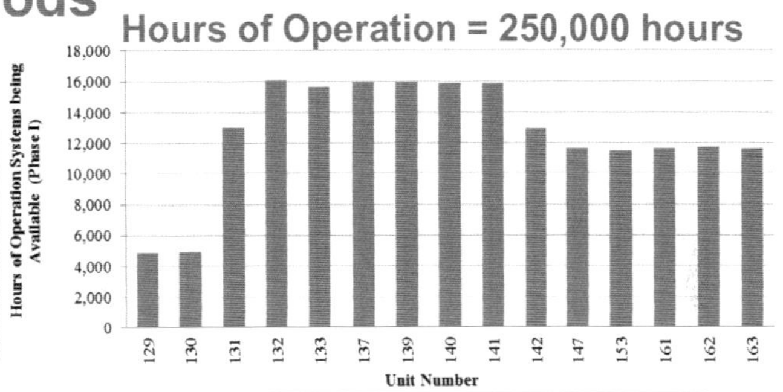

- Initial Operation
  - Oct 2011-Feb 2012
  - Stack degradation
  - Could not maintain 5 kWe

- Set Point Change
  - Mar 2012- Jun 2012
  - Decreased setpoint to 4 kWe
  - Availability dropped 95.7% → 89%

- BOP Upgrades
  - July 2012 to Shutdown
  - Parts upgraded:
    - Steam Pump Assembly (SPA)
    - New air flow sensor and Oil Pump Assembly (OPA)
    - Improved water filtration
    - Improved igniter
- Availability increased from 89% to 96.5%

Engineering Results from Phase I

Electrical Power

4.1 kWe

Heat Produced

4.6 kWt

Availability

93.5%

Water Temperature

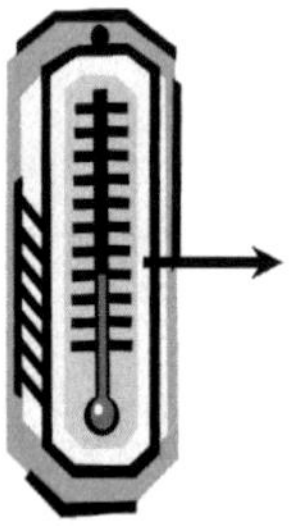

123° F

System Efficiency (HHV)

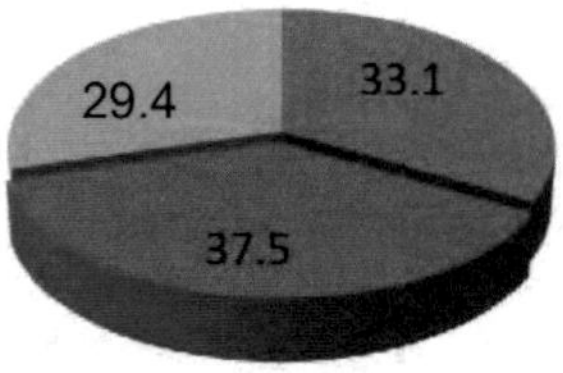

70.6% Total Efficiency

- Electric Efficiency
- Heat Recovery Efficiency
- Losses

11

Electrical Utilization

- 100% utilization of electricity

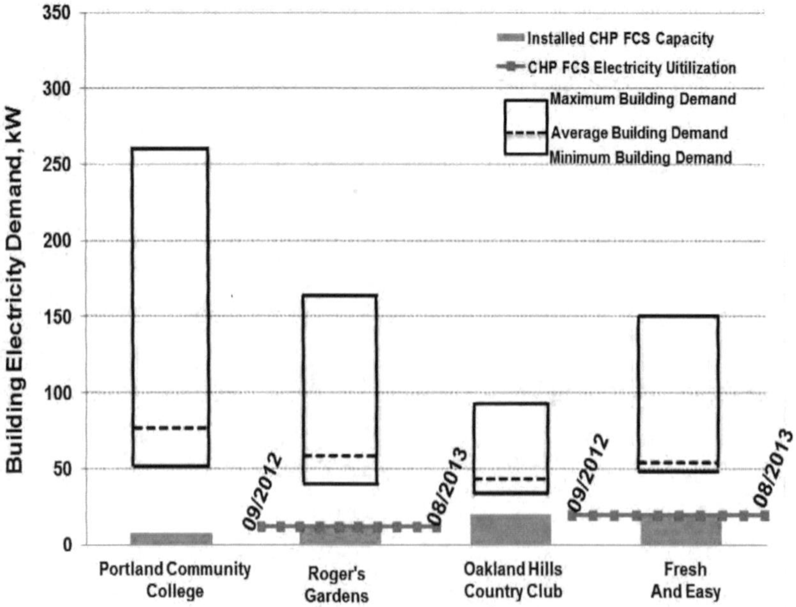

12

Heat Utilization

- Roger's Garden: 0% utilization
  - Instrumentation moved to Oakland Hills Tennis Club
- Fresh and Easy: 90% utilization (Nov 2012-Aug 2013)

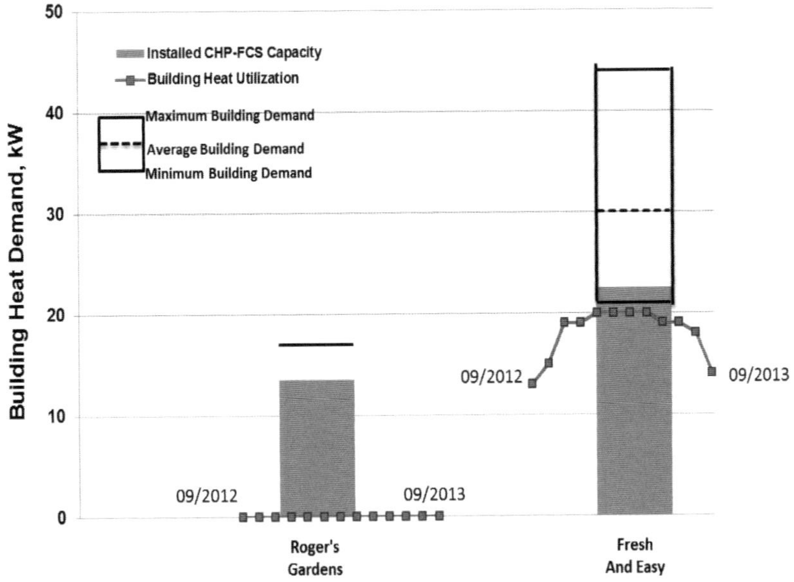

13

CE5 Failures

- Significant number of non-part failures create downtime
- Slight decrease in premature part failures over time

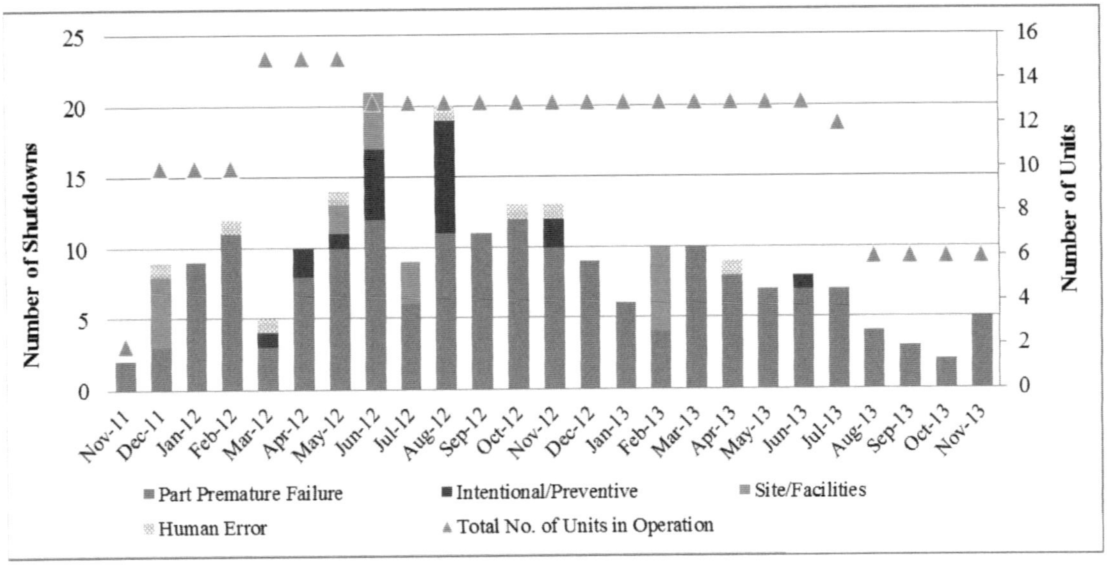

Phase II Performance

- ClearEdge upgrades CE5 to PureCell System Model 5 (M5) at no cost to DOE
  - 2 units in July/Aug. 2013
  - 8 units in Jan./Feb. 2014

Hours of Operation = 44,000 hours

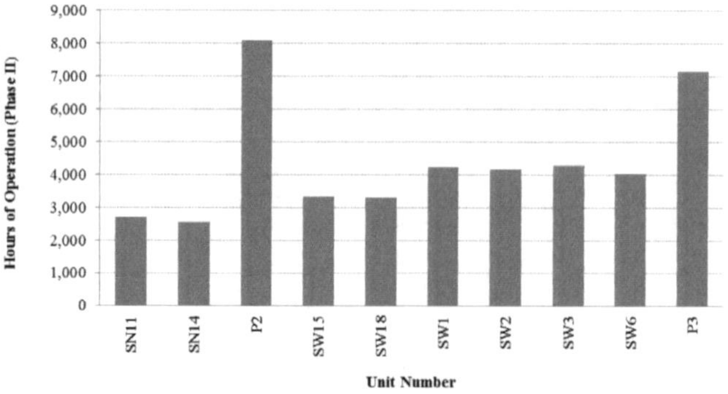

- P2 operates for > 8000 hours with 99.7% availability
- P3 operates for > 7000 hours with 91.7% availability
- Average availability 94.8%

- May 2014 ClearEdge files for bankruptcy
- July 2014 Doosan purchases ClearEdge and discontinues data collection on all M5 units
- All systems still operating

Pacific Northwest NATIONAL LABORATORY

Proudly Operated by Battelle *Since 1965*

15

CE5/M5 Performance Comparison

- Most significant difference is stack performance as evidenced by overall efficiency
- Standard deviation of overall system efficiency for one of the CE5 units reduced from 5.8 to 0.7

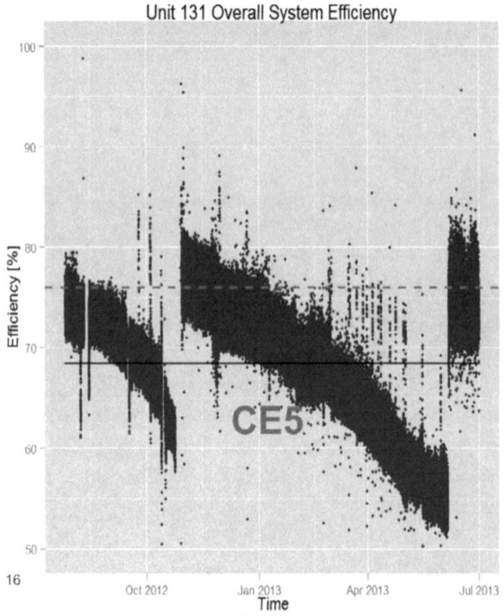

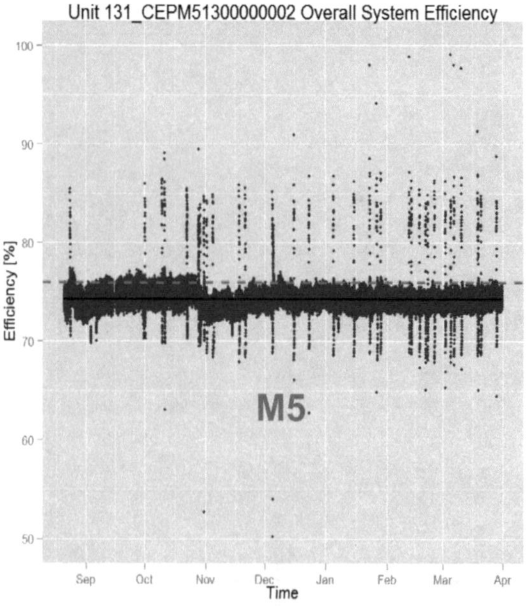

16

Power and Efficiency Distribution

- Unit 131 and P2 one year comparison
- Much larger standard deviation for CE5 compared to M5

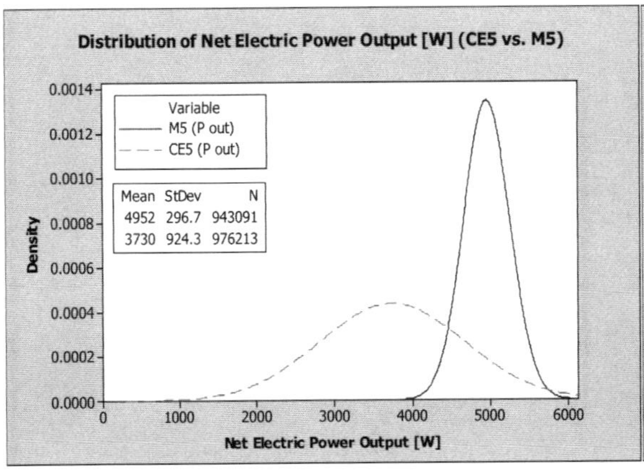

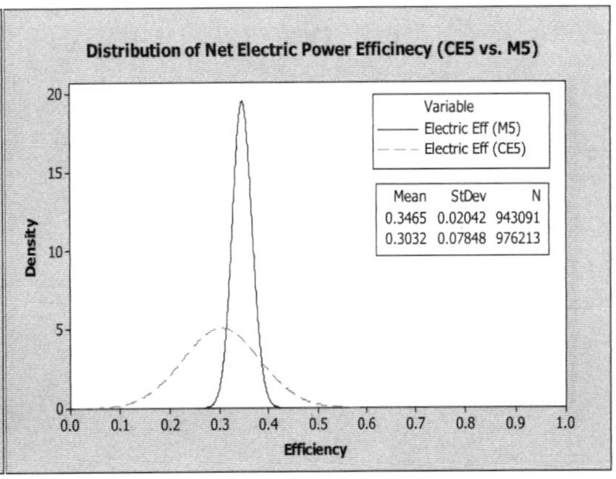

Electrical and heat utilization data collection not performed in Phase II

Pacific Northwest
NATIONAL LABORATORY

Proudly Operated by Battelle *Since 1965*

17

Comparison of Engineering Results

Electrical Power

4.1 → 4.8 kWe

Heat Produced

4.6 → 5.4 kWt

Availability

93.5% → 94.8%

Water Temperature

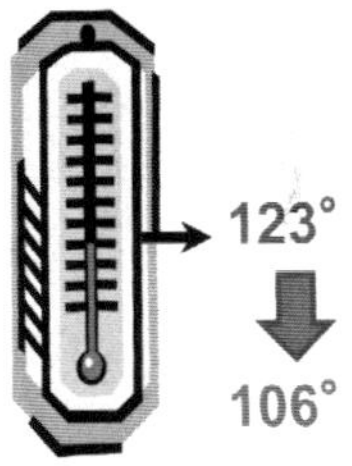

123° F → 106° F

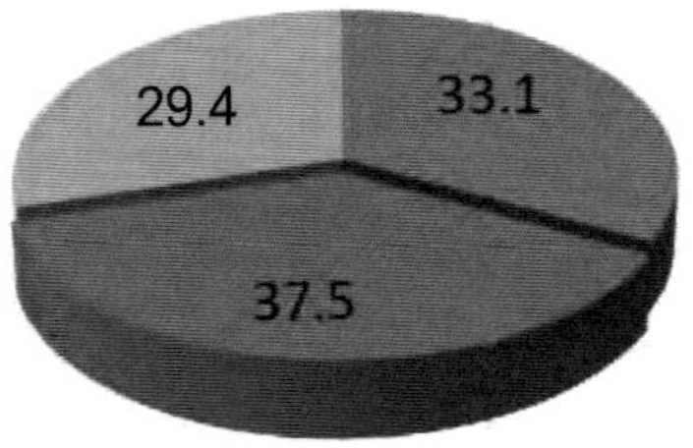

CE5

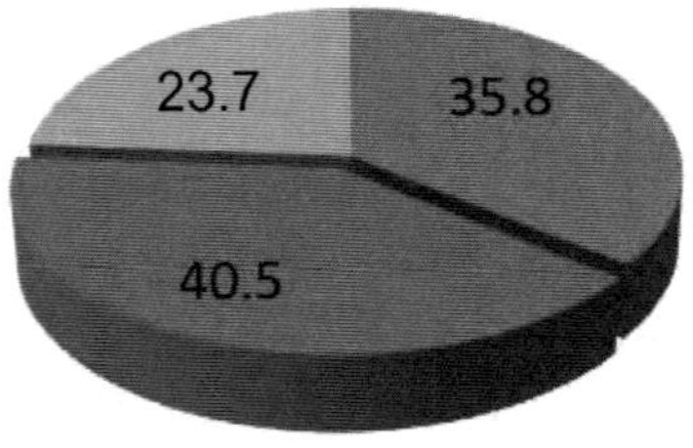

M5

- Electric Efficiency
- Heat Recovery Efficiency
- Losses

Failure Comparison between CE5 and M5

- No M5 non-part failures
- 0.80 shutdowns/month/unit for CE5 units
- 0.46 shutdowns/month/unit for M5 units
- 42% reduction in number of maintenances required

Unit Number	Part Premature Failure		Intentional / Preventive		Site/Facilities		Human Error	
	CE5	M5	CE5	M5	CE5	M5	CE5	M5
129-SN11	5	5	0	0	1	0		0
130-SN14	4	0	0	0	1	0		0
131-P2	13	2	3	0	1	0	1	0
132-SW15	20	2	4	0	0	0		0
133-SW18	24	1	4	0	0	0		0
137-SW1	12	4	0	0	2	0		0
139-SW2	15	3	0	0	3	0	1	0
140-SW3	11	0	0	0	2	0		0
141-SW6	15	2	1	0	3	0		0
142-P3	10	13	0	0	3	0	1	0
147	12	0	1	0	0	0		0
153	14	0	1	0	1	0	1	0
161	9	0	2	0	1	0	2	0
162	5	0	2	0	1	0		0
163	16	0	2	0	1	0	2	0

19

Comparison to DOE Targets for 1-10 kWe CHP-FCS Operating on Natural Gas

- Total Energy Efficiency
  - Total energy efficiency improved by about 5.6% with the CE5 to M5 upgrade.
  - Still need 11.2% system efficiency increase to meet 2015 technical target
- Availability
  - Availability increased by 1.3% with the CE5 to M5 upgrade
  - Still need 3% system availability increase to meet 2015 technical target

Performance Indicators	2015	CE5 (Average)	M5 (Average)
Electrical Efficiency at Rated Power	42.5%	33.1%	35.8%
CHP Energy Efficiency	87.5%	70.7%	76.3%
Operating Lifetime	40,000 h	40,000 (Expected)	80,000 (Expected)
System availability	98%	93.5%	94.8%

20 Proudly Operated by Battelle Since 1965

Conclusions

- Comparison between CE5 and M5 shows higher performance and reliability
  - M5 unit was degraded 88% less than its peer CE5 unit
  - 0.8 shutdowns/month for CE5 systems and 0.46 for M5 units
  - M5 has 3% higher electrical and thermal efficiency than CE5
- Significant degradation with PBI fuel cell (CE5) but no degradation seen with phosphoric acid fuel cell (M5) over one year of operation
- Project demonstrated significant improvement in fuel cell technology over 3 years of evaluation
  - Improved stack technology
  - Addressed BOP reliability

21

Acknowledgements

- This research was supported by the U.S. Department of Energy (DOE) Office of Energy Efficiency and Renewable Energy (EERE) Fuel-Cell Technologies (FCT) Program.
- The authors wish to thank Peter Devlin of the U.S. DOE EERE FCT, for his support and direction.
- The authors also would like to acknowledge Dr. Siva Pilli for his technical support.

U.S. Department of Energy
Energy Efficiency and Renewable Energy
Bringing you a prosperous future where energy is clean, abundant, reliable, and affordable

Pacific Northwest
NATIONAL LABORATORY

Proudly Operated by Battelle *Since 1965*

22

Questions?

Contact:

Atefe Makhmalbaf ☎: 509-372-6784 ✉: atefe.makhmalbaf@pnnl.gov

Kriston P Brooks ☎: 509-372-4343 ✉: kriston.brooks@pnnl.gov

Pacific Northwest
NATIONAL LABORATORY

Proudly Operated by Battelle *Since 1965*

23

366125 Revisiting the Application Niche of Microbial Fuel Cell Technology for Sustainable Wastewater Treatment

Abstract:

Microbial fuel cells (MFCs) are an emerging concept for sustainable treatment of wastewater with simultaneous recovery of bioenergy. In the past decade, MFCs have been intenstively studied and much knowledge has been obtained from the aspects of microbiology, materials, electrochemistry, and reactor configuration and operation. A great challenge in MFC development is how to use fundamental knowledge obtained from laboratory studies to aid its further development. A key task is to identify proper application niche for MFC technology during wastewater treatment. For many years, the major function of MFCs has been considered as "bioenergy production". The research findings in the recent years reveal that the energy conversion efficiency with actual wastewater in an MFC is low, and it may not be reasonable to treat MFCs as a major energy producer. However, MFCs may still possess some advantages that can make it potentially useful under a certain condition. Understanding how MFCs can be potentially applied will be critical to allocate the limited research resource for efficient development of this technology. In this presentation, we will rethink MFC applications and discuss the challenges and perspectives. We will propose several application niches based on the research findings that may accelerate MFC research towards a practical technology.

373879 Hydrophobic and Hydrophilic Membrane for DCMD: Effect of Non-Solvent Additives and Operation Condition on the Performance

Abstract:

Dual-layer composite membrane is a new design for direct contact membrane distillation (DCMD) with membrane performance potentially superior to that of single layer porous membranes. Using dry-wet phase inversion technology, novel dual-layer hollow fiber membranes were fabricated in current research. The outer layer was made from polyvinylidene fluoride (PVDF) with polyvinylpyrolidone (PVP), glycerol or ethanol as non-solvent additive, while the inner layer consists of PVDF and polyvinyl alcohol (PVA) blend. Effects of PVA/PVDF blending ratio and non-solvent additive type on the morphological, mechanical and separation characteristics of the composite membranes were investigated. It was found that increasing the PVA content has no obvious effect on the macrovoids formation, but increase the pore size located inside the spongy section. The DCMD tests showed that the change in the composition of inner layer material led to the variation of the flux and rejection coefficient. The higher PVA content gave rise to higher hydrophilicity of inner layer, and hence better wettability by water, which was confirmed by elementary analysis. But enhancing the PVA content in the inner layer was related with lower flux and lower rejection. Considering the higher thermal conductivity coefficient of water, the decreasing trend of flux with improved inner layer hydrophilicity was attributed to the transport of water trapped inside the hydrophilic porous structure. The lower rejection was probably due to the relatively hydrophilic interface serving as the distillate side of the hydrophobic layer. It was also found that the pattern of the relation between flux/rejection and liquid flow rate were different between lumen side and shell side. The findings in current work suggests: 1) the surface energy of the inner layer is important to the performance of the dual-layer hydrophobic/hydrophilic DCMD membrane; 2) the transport resistance in the inner layer should not be neglected.

The treatment of highly concentrated wastewater consisting of two reactive dyes, two acid dyes and one basic dye was studied by individual coagulation/flocculation (CF) and nanofiltration (NF) processes as well as their combination (referred as CF-NF). Polyaluminum chloride (PAC) and polydiallyldimethyl ammonium chloride (PDDA) were screened out and selected as the most effective coagulant and flocculant, respectively. It is found that the CF process are favorable for highly concentrated dye solutions with pH values large than 3. It can achieve about 90% of dye removal at the optimal dosage of PAC/PDDA=400/200 ppm. A positively charged NF hollow fiber membrane was fabricated and used for NF treatment. It is able to remove almost 100% dyes with a permeate flux of about 1.0 $Lm^{-2}h^{-1}$ under an operating pressure of 1 bar. The combination of CF and NF can complement each other's strengths and overcome their individual limitations. The NF treatment can completely remove the strong color left in CF treated dye solutions, while the efficiency of coagulant/flocculant is improved by treating NF concentrated streams and subsequently results in much less sludge. In addition, membrane fouling is abated and NF permeate flux is increased by applying the CF process as a pretreatment. Thus, combining CF and NF together improves the overall performance for wastewater treatment.

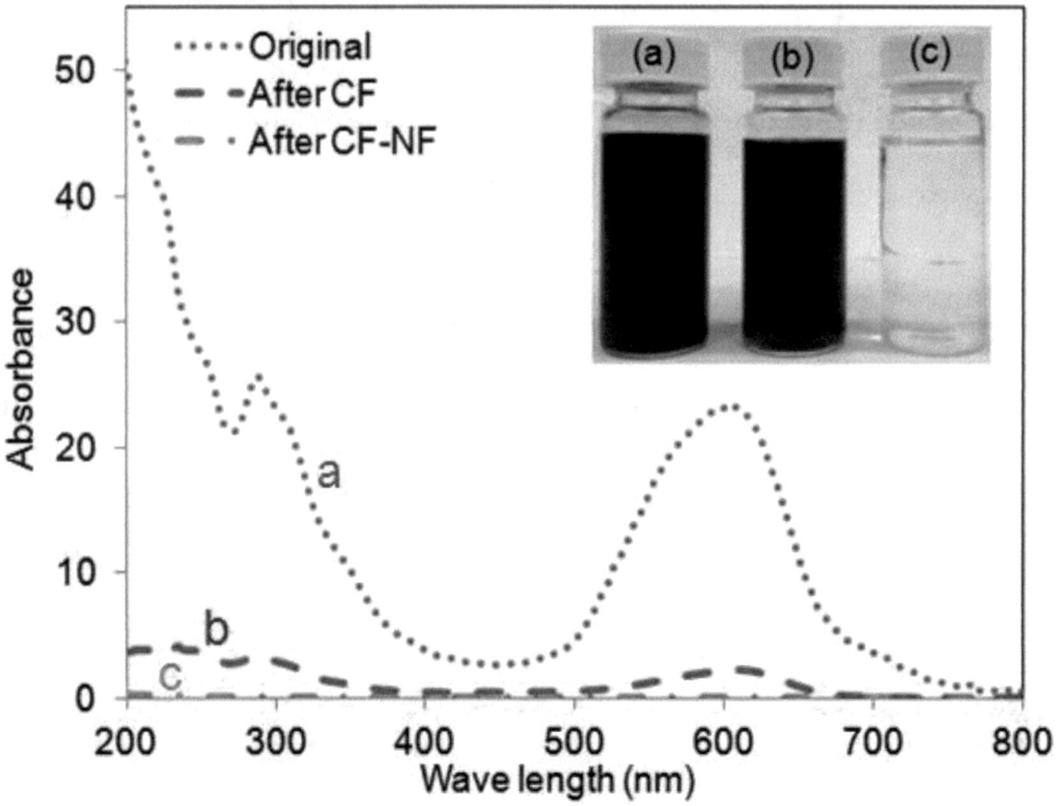
Original
After CF
After CF-NF
(a)
(b)
(c)
a
b
c
Absorbance
Wave length (nm)
0
10
20
30
40
50
200
300
400
500
600
700
800

374213 Synthesis and Application of Multi-Carboxyl Compound As a Novel Draw Solute in Forward Osmosis Process

Abstract:

The forward osmosis (FO) process has attracted growing research interest as an alternative water-treatment technology for its low energy consumption and high contaminant rejection. However, the limited choice of suitable draw solution is still an obstacle in the commercial development of FO technology. In this work, a multi-carboxyl compound-ethylenediaminetetrapropanoic acid (EDTP acid) has been synthesized via the Michael addition reaction and employed as a novel draw solute in FO process. The as-synthesized structure is confirmed by FTIR, NMR, and HRMS. By proper optimizing the pH of EDTP acid solution, the solubility and the osmotic pressure can be effectively improved. The FO performance with EDTP acid/salts as draw solutes is evaluated, and the results suggest that water flux and salt flux both increase with the increasing solute concentration. A high water flux of 22.36 LMH could be achieved with 0.8mol/L EDTP solution as draw solution and deionized water as feed solution through a HTI-CTA FO membrane in the pressure-retarded osmosis mode. Clearly, EDTP acid is superior to most traditional draw solutes for its much less draw solute leakage with a comparable water flux in the FO process. The pressure-driven NF process was also employed to regenerate the diluted draw solution effectively.

386411 Microbial Fuel Cell for Harnessing Sustainable Energy from Wastewaters Containing High Salt Concentrations or Produced Water

Abstract:

Production of natural gas via fracking generates wastewater causing environmental pollution threatening the health of our drinking water supplies, rivers, streams and groundwater. The process generates massive amounts of polluted produced water. Produced water is conventionally treated through different physical, chemical, and biological methods. However, the conventional technologies cannot remove dissolved elements effectively. An upcoming technology, microbial fuel cells (MFC), can be a potential solution. It can generate bioenergy from the organics in the produced water. MFC application has documented a significant interest in the contemporary research area in harvesting energy in the form of bioelectricity. Anodic bacteria of the MFC play critical role in power generation and treatment efficiency of MFC. One of the approach to enhance the efficiency of MFC is enrichment of electroactive microbial biofilms capable of tolerating high salt concentrations. In current study, the effect of salt concentration in the anode chamber of a MFC fed with sodium acetate was tested with increasing concentration of NaCl from 10 g/L to 45 g/L. An MFC design with low electrode spacing, high specific electrode surface area with minimal dead volume, air-cathode and control of external resistance was used. Maximum current density of 7 A/m2 was observed with the salt concentration of 35 g/L. The effect of salt concentration on current density and columbic efficiency will be discussed in light of the sensitive nature of the anodophilic bacteria to high salt concentrations.

<u>**Options for improving attractiveness of renewable energy powered desalination**</u>

Hiren D. Raval[a]*, Subarna Maiti[a], Abbas Ghassemi[b], Leila Karimi[b]

**Reverse Osmosis Discipline,*
[a]CSIR-Central Salt and Marine Chemicals Research Institute (CSIR-CSMCRI),
Council of Scientific & Industrial Research (CSIR),
Gijubhai Badheka Marg,
Bhavnagar- 364 002, (Gujarat), INDIA
Fax: +91-0278-2566970
E-mail: hirenraval@csmcri.org

[b]Institute for Energy and the Environment/WERC
New Mexico State University
1060 Frenger Mall, ECIII, Suite 300 S
PO Box 30001 MSC WERC
Las Cruces, NM 88003-8001 USA

Abstract: With an advancement of technology and rampant growth in installed desalination capacity in last decade, it has become imperative to analyze the energy required for the same and associated greenhouse gas emission. With an increasing concern over the green house gas emission, the attention is being paid to the area of renewable energy. More recently, the thrust is being given to renewable energy driven desalination plants, especially powered by solar energy. Solar powered desalination is also being considered as an option for remote and sparsely populated locations, not connected with electrical grid. The present paper discusses the concept of overall energy efficiency.

Life cycle green house gas emission of a coal based thermal power plant is 1001 g CO_2/ KWH whereas the same for solar PV is 46 g CO_2/KWH. Considering the specific power consumption of 4 KWH per cubic meter of water produced and approximately 10% of global desalination capacity being converted to solar powered desalination i.e. 7840 million liters per day, close to 29,950 tons of CO_2 will be saved from getting emitted in world's environment, each day. This justifies the research on renewable energy powered desalination.

Renewable energy powered desalination seems very attractive on prima facie; however, the high capital cost of renewable energy generation and lower reliability in terms of consistency are the major impediments in its implementation. To make the renewable energy powered desalination attractive, the energy requirement for desalination should be minimized and the options of using low grade energy i.e. waste heat can be explored.

The options for reducing the power consumption by reverse osmosis have been explored world over. There has been thrust on developing the technology of nano-composite membrane that reduces the energy consumption by ca. 20% by imparting nano-material in the barrier layer of TFC membrane. There has also been thrust in perfecting the technology of recovering the energy from the concentrate stream by using different types of energy recovery devices e.g. pressure exchangers. However, there is a little focus on very easy option of reducing the energy consumption by increasing the feed water temperature. The feed water temperature can be increased by any low grade heat source. Increased temperature of feed water improves the product water flow rate to a substantial extent with slight decline membrane selectivity. This will be particularly useful for brackish water reverse osmosis where the re-mineralization of product water can be avoided by slightly lower selectivity of the membrane and improved flow rate results in decrease of energy consumption to a greater extent. The synergy can be achieved when the high flux membrane is used with high temperature feed water to significantly reduce the power consumption.

Hydraulic permeability increases at higher temperature which in turn results in improved ϒ. Similarly, increasing number of stages and using energy recovery devices will also reduce the specific energy consumption for reverse osmosis.

In this way, the present paper analyzes the options for reducing specific energy consumption to make renewable energy powered desalination attractive and also demonstrates the experimental results to improve the performance of photovoltaic panel and also to increase the overall energy efficiency.

Introduction

With an advancement of technology and rampant growth in installed desalination capacity in last decade, it has become imperative to analyze the energy required for the same and associated greenhouse gas emission. With an increasing concern over the green house gas emission, the attention is being paid to the area of renewable energy. Solar powered reverse osmosis has been studied the most extensively out of all renewable energy powered desalination systems. Owing to the ease of operation and modular set up, photovoltaic powered reverse osmosis has gained popularity. Solar powered desalination is also being considered as an option for remote and sparsely populated locations, not connected with electrical grid. The present paper discusses the concept of overall energy efficiency.

Many researchers installed the experimental facility to power reverse osmosis plants with solar photovoltaic technology. However, cost of producing fresh water by such a process is rather expensive- reported to be 6.52 $/cubic meter of fresh water over the 20- year life time of the equipment. [1] A prototype photovoltaic-powered reverse-osmosis system has been reported where the rate of production of fresh water varies throughout the day according to the available solar power, and thus, the system operates without need of batteries. The system is designed to operate from seawater, and a Clark pump brine-stream energy recovery mechanism is coupled with a variable recovery ratio technique to achieve a specific energy consumption of less than 4 kWh/m^3 over a wide range of operation. [2]

The solar thermal-powered reverse osmosis desalination system was coupled to a solar power cycle based on a Rankine cycle with toluene, hexamethyl disiloxane (MM) and octamethyl cyclotetrasiloxane (D4) as working fluids and two different models of parabolic trough collectors. [3] A solar-powered trans-critical CO_2 (carbon dioxide) power cycle for reverse osmosis desalination based on the recovery of cryogenic energy of LNG (liquefied natural gas) has been studied. [4]

A hybrid wind/solar powered reverse osmosis desalination system has been modelled and simulated. [5] A solar thermal and photovoltaic-powered reverse osmosis (RO) desalination plant has been constructed and optimized for brackish water desalination. [6] Desalination of brackish water as a viable option to cope with water scarcity and to overcome water deficit in Jordan is assessed. [7] A prototype photovoltaic-powered reverse-osmosis system has been constructed at CREST, Loughborough, UK. The rate of production of fresh water varies throughout the day according to the available solar power, and thus the unit operated without batteries has been demonstrated. [8] The novel solar powered direct osmosis desalination process has been demonstrated. [9] The life cycle Greenhouse Gas (GHG) emissions of a Seawater Reverse Osmosis (SWRO) desalination plant has been assessed and found that GHG emissions reduction of ~90% can be achieved by opting for renewable energy. [10]

Thus, many researchers attempted to study solar powered reverse osmosis and its application for seawater and brackish water desalination. However, the potential of tapping thermal energy from solar photovoltaic panel is attractive and can be coupled with desalination. The present paper demonstrates the improvement in overall energy efficiency by this approach.

Rationale: Life cycle green house gas emission of a coal based thermal power plant is 1001 g CO_2/ KWH whereas the same for solar PV is 46 g CO_2/KWH [11]. Considering the specific power consumption of 4 KWH per cubic meter of water produced and approximately 10% of global desalination capacity being converted to solar powered desalination i.e. 7840 million liters per day [12], close to 29,950 tons of CO_2 will be saved from getting emitted in world's environment, each day. This justifies the research on renewable energy powered desalination.

Renewable energy powered desalination seems very attractive on prima facie; however, the high capital cost of renewable energy generation and lower reliability in terms of consistency are the major impediments in its implementation. To make the renewable energy powered desalination attractive, the energy requirement for desalination should be minimized and the options of using low grade energy i.e. waste heat can be explored.

The options for reducing the power consumption by reverse osmosis have been explored world over. There has been thrust on developing the technology of nano-composite membrane that reduces the energy consumption by imparting nano-material in the barrier layer of TFC membrane. There has also been thrust in perfecting the technology of recovering the energy from the concentrate stream by using different types of energy recovery devices e.g. pressure exchangers. However, there is a little focus on very easy option of reducing the energy consumption by increasing the feed water temperature. The feed water temperature can be increased by any low grade heat source. Increased temperature of feed water improves the product water flow rate to a substantial extent with slight decline membrane selectivity. This will be particularly useful for brackish water reverse osmosis where the re-mineralization of product water can be avoided by slightly lower selectivity of the membrane and improved flow rate results in decrease of energy consumption to a greater extent. The synergy can be achieved when the high flux membrane is used with high temperature feed water to significantly reduce the power consumption.

The specific energy consumption can be reduced by the following methods from the first principle. [13]

1. Increasing $\Upsilon = A_{total} Lp \Delta\pi_o / Q_f$ ....(1)
2. Increasing number of stages
3. Using energy recovery device

   $Lp = C_{LP} \bullet \exp(-Ea^{LP}/RT)$ [14] .... (2)

   Where Lp = hydraulic permeability

   C_{LP} = Constant

   Ea^{LP} = Activation energy represents the per mole difference in enthalpy of a molecule which is necessary to overcome the transport barriers during its passage across the membrane

   T = Temperature

   Substituting (2) in (1)

   $$\Upsilon = A_{total}\, C_{LP} \bullet \exp(-Ea^{LP}/RT)\, \Delta\pi_o / Q_f \text{ .... (3)}$$

Therefore, T has to be maximized for maximizing ϒ. Hydraulic permeability increases at higher temperature which in turn results in improved ϒ. Similarly, increasing number of stages and using energy recovery devices will also reduce the specific energy consumption for reverse osmosis.

Tapping thermal energy from the low grade energy source e.g. solar photovoltaic panel can increase the feed water temperature for reverse osmosis. It has been demonstrated that the thermal energy can be tapped from solar photovoltaic panel to increase the overall energy efficiency of the PV panel. [15]

Experimental

Heat transfer from solar PV panel will increase the panel efficiency and also tap the thermal energy, that is goes unutilized otherwise. The direct contact heat exchanger system was designed with the coolant being water. Since the radiations incident from top, the top panel surface heat exchange can effectively control the temperature of PV panel. The sides and back of the panel was properly insulated with calcium silicate insulation to abate heat losses. Rationale was to utilize the maximum thermal energy and minimize the losses.

Materials: Solar PV panel - 70 Wp (M/s Kotak Urja Limited), frame structure, water tank, Rheostat, K type thermocouples, , pyranometer (Kipp & Zonen CM4 pyranometer).

Method:

As shown in figure 1, the PV panel was kept at 20° inclination in southward direction to get the optimal access to solar radiation with reference to the location Bhavnagar, India, Co-ordinates: 21.7600^o N and 72.1500^o E. One PV panel was provided cooling from top, whereas the other panel was kept without cooling. The variable resistance system (Rheostat) was used to measure the V-I (Voltage- Current) performance of PV panel.

The system where cooling was provided from top comprised of the perforated pipe, perforations being 10 mm in diameter. The flow rate of cooling water was varied from 1 liter per minute to 2 liter per minute and the V-I performance of the PV panel was evaluated. The water at outlet was drained out in a tank open to atmosphere and was then recirculated using a DC (direct current) Kemflo make pump. The nominal flow rate of the pump was 1 $Lmin^{-1}$. Two pumps are operated for getting flow of 2 $Lmin^{-1}$.

Figure 1: Water flowing from top of the solar photovoltaic panel

Modelling:

The ANSYS Computational fluid dynamics software was used to simulate this model. Assumptions made in simulating the model using ANSYS CFD tool are depicted below.

- The flow is steady
- A constant water thickness of 2 mm above the panel is considered.
- The water temperature varies with time.
- Air temp and density also vary with time.
- Visible radiation: 40%, IR : 50% and ultraviolet : 10% of the total incident radiation.
- Glass transmissivity :80% absorptivity :20% and no reflectivity
- Absorptivity of opaque material is 80%
- Pipe material does not absorb radiation.

The photovoltaic panel comprised of the following different layers, physical properties of each layer is given in table 1.

Table 1: Physical properties of the constituents of PV panel

Layer	**Thickness t (m)**	**Thermal Conductivity K (w/m K)**	**Density (kg/m^3)**	**Specific heat Cp (J/kg K)**
Glass	0.003	1.8	3000	500
ARC	100X10^{-9}	32	2400	691
PV Cell	225X10^{-6}	148	2330	677
EVA	500X10^{-6}	0.35	960	2090
Rear contact	10X10^{-6}	237	2700	900
Tedlar	0.0001	0.2	1200	1250

The geometry was created with air domain in ANSYS CFD tool and meshed. The meshed domain is shown in figure 2.

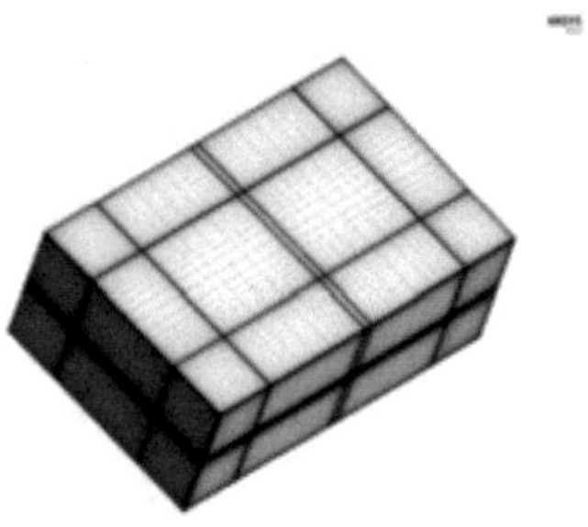

Fig. 2: Meshed domain

The meshing characteristics: Skewness: 0, Orthogonal quality: 1

The snapshot in figure 3A and 3B indicates the comparison of PV panel temperature with 1 LPM and 2 LPM flow respectively at 1000 Hrs and 1300 Hrs. It is evident from simulation results that larger area of PV panel remains at lower temperature with increased flow rate, and higher flow rate is particularly required to maintain the temperature when solar radiation intensity is higher at 1300 Hrs. While comparing the panel temperature without cooling as shown in figure 4, it becomes evident that cooling is very efficient in controlling the panel temperature. The panel temperature without cooling reaches upto 76°C and changes at different locations in the panel as a result of natural convection as indicated in figure 4. The panel temperature is controlled within 40°C by cooling even with lower flow rate of 1 LPM.

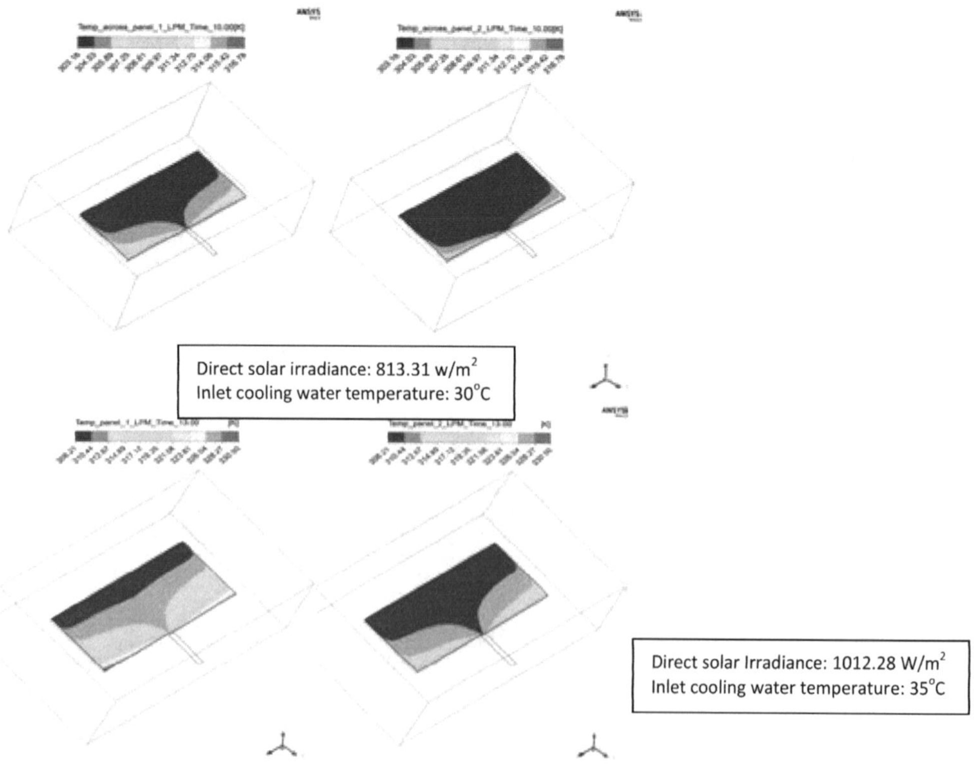

Figure 3A (Top) and 3B (Bottom): PV panel temperature with cooling- 1LPM Flow (1000 Hrs) and 2 LPM flow (1300 Hrs) respectively

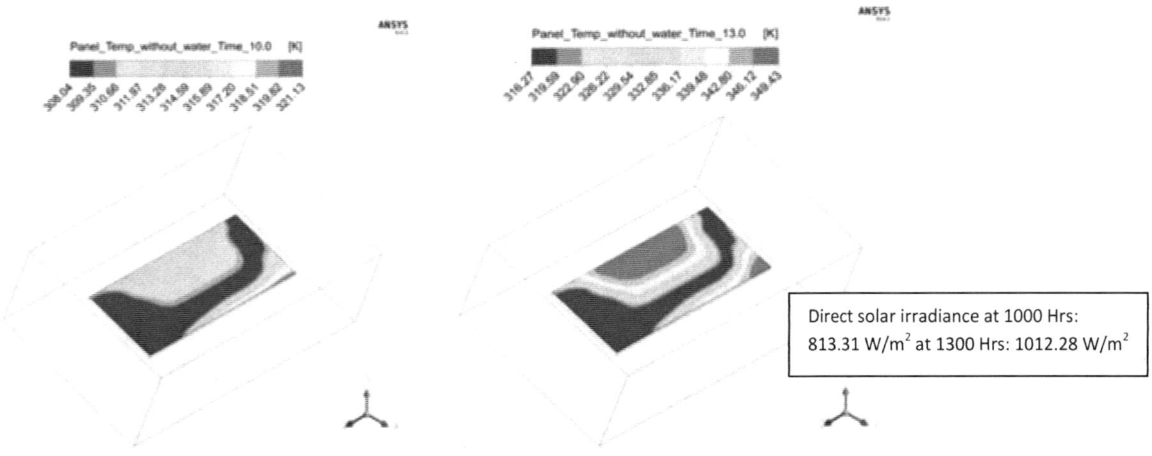

Figure 4: Panel temperature without cooling 1000 Hrs and 1300 Hrs

Results and Discussions

The ambient conditions on the days of experiment are as shown in table 2.

Table 2: The environmental conditions on the days of experiment

Condition	14 May	1 June
Insolation (7 Hrs- 1030 to 1730)	837.3655 W/m^2	838.709 W/m^2
Average ambient temperature	36.5 ^{o}C	38.62 ^{o}C
Average wind speed	1.43 m/s	1.4 m/s
Average insolation (24 hrs)	488.6 W/m^2	343.84 W/m^2

The performance of PV panel is assessed by V-I Performance. In figure 5, the performance of PV panel has been demonstrated, where the peak power produced by PV panel with and without cooling are compared. It is clear that the peak power produced by PV panel improves with cooling where about 10% improvement in power output is observed at 1300 hrs. The pump consumes 5W power. Therefore, the net power produced with cooling has to be reduced by 5W in each case. However, the gravity operated systems can also be designed where; net power will be the same as power produced.

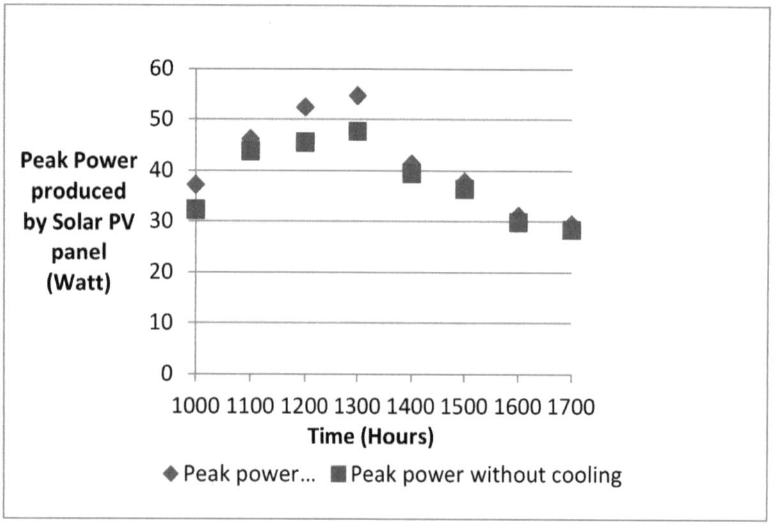

Figure 5: Performance comparison of PV panel with and without cooling – Cooling water flow: 1 LPM

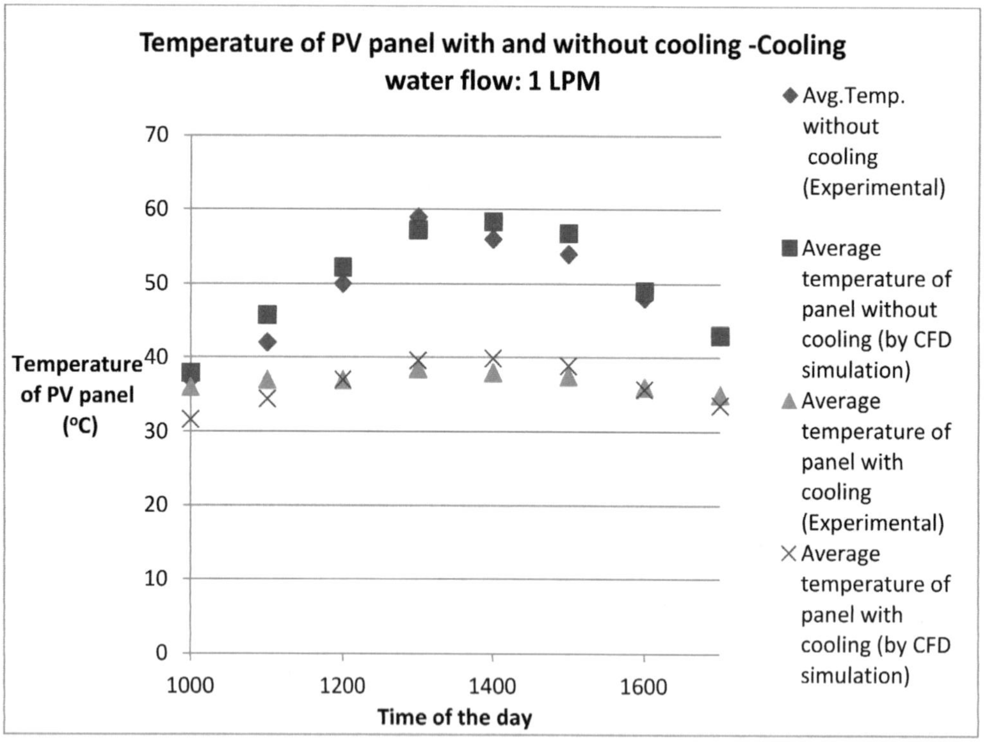

Figure 6: Temperature of PV panel with and without cooling

It is explicit from figure 6 that there is a significant decline in PV panel temperature as a result of cooling from top surface at 1 LPM flow. The experimental results are compared with CFD simulation results and found in close conformity. When the temperature reaches close to 60° C at 1400 hrs without cooling the panel, it is controlled well below 40° C with cooling.

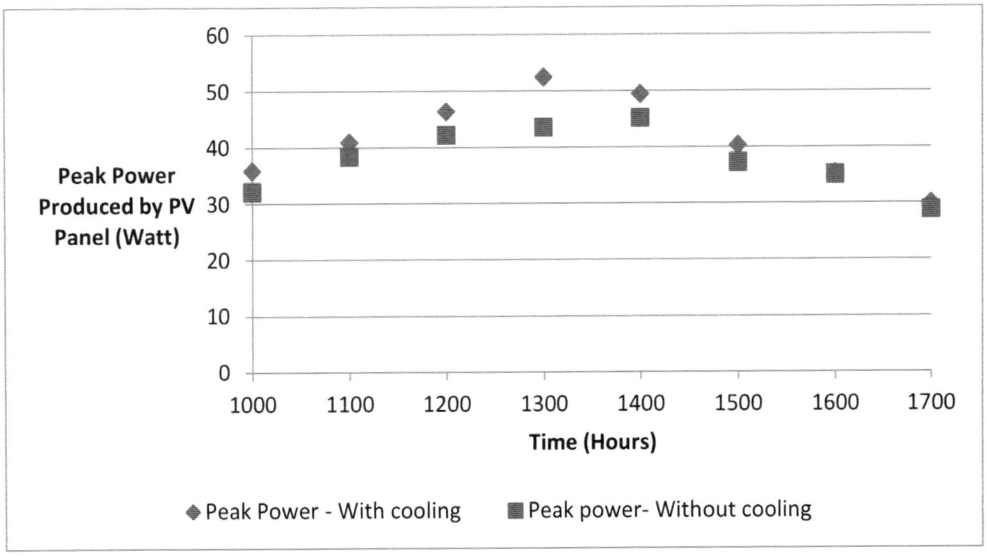

Figure 7: Performance comparison of PV panel with cooling water flow-2 LPM and without cooling

It is evident from the figure 7 that the peak power produced by PV panel improves as in the case of 1 LPM flow, however the improvement is substantial. There is 20% improvement in peak power produced at 1300 hrs which is higher than that with 1 LPM flow. The pump consumes 5 W power. Therefore, the net power produced with cooling has to be reduced by 5 W in each case. However, the gravity operated systems can also be designed where; net power will be the same as power produced as mentioned above.

Refraction effect and improvement of panel performance

The potential for direct liquid-immersion cooling for improving CPV cell electrical performance has been demonstrated. However, the influence of refractive index in changing direction of the solar radiation has not been studied.

The reason of improvement in energy efficiency can also be attributed to the refraction effect because of water flow above the panel. The angle of incident solar radiation in Bhavnagar was obtained by SOLPOS calculator. The angle changed as a result of refraction by the water layer. The refractive index of air and water are known. The incident angle is known. The angle after refraction can be calculated using Snell's law $n_1 \sin\theta_1 = n_2 \sin\theta_2$.

The table 3 indicates the effect of refraction over the solar PV panel.

Table 3: The Angle change as an effect of refraction by solar PV panel

Time	AngleΘ_1	n_1	n_2	sinΘ_1	SinΘ_2	Θ_2
900	-61.91	1.00	1.33	-0.88	-0.66	-41.58
1000	-50.74	1.00	1.33	-0.77	-0.58	-35.63
1100	-41.79	1.00	1.33	-0.67	-0.50	-30.09
1200	-36.76	1.00	1.33	-0.60	-0.45	-26.76
1300	37.26	1.00	1.33	0.61	0.46	27.10
1400	43.10	1.00	1.33	0.68	0.51	30.94
1500	52.53	1.00	1.33	0.79	0.60	36.66
1600	63.97	1.00	1.33	0.90	0.68	42.53
1700	76.49	1.00	1.33	0.97	0.73	47.00

From the table 3, it is clear that the span of angle varied from -61.91 to 76.49 from Morning 0900 hrs to evening 1700 hrs. As a result of refraction the angle θ_2 varied from -41.58 to 47.00. In this way, the span of angle reduced from 138.4 to 88.58. In other words, the incident radiations were straightened as a result of refraction as shown in the figure 8. These straightened radiations helped to improve the performance of the panel as they strike to panel at relatively larger angle as compared to the one without water layer on top.

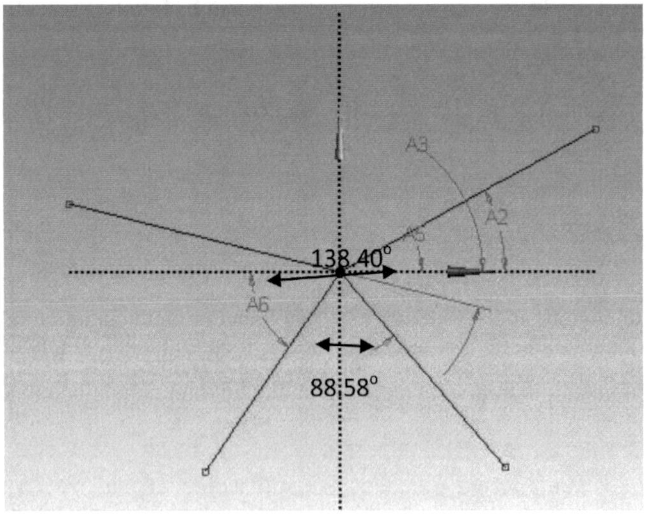

Figure 8: Radiation at a given point straightened as a result of refraction

Overall energy efficiency: The energy efficiency with and without cooling was worked out as a case to understand the effect of cooling on overall energy efficiency. It has been observed that the efficiency was raised from 6.68% to 40.42% with cooling as shown in table 4.

Table 4: Energy efficiency calculations (For 1 Liter per minute flow)

Time	1200 Hrs	1400 Hrs	1600 Hrs
Inlet water temperature(Tw_{in})	37	36	34
Outlet water temperature(Tw_{out})	39	38	35
Energy contained by water(watt)	139.57	139.57	69.78
Peak power without cooling (watt)	34.50	25.28	14.22
Peak power with cooling (watt)	41.72	28.56	16.3
Difference (watt)	7.22	3.28	2.08
Total power saving (watt)	146.79	142.85	71.86
Watt/m^2	1020	732	348
Solar panel area (sq.m)	0.61	0.61	0.61
Power incident on the panel (watt)	624.24	447.98	212.98
Energy efficiency without cooling	5.53	5.64	6.68
Energy efficiency with cooling	29.04	37.53	40.42

Conclusion

The photovoltaic panel temperature increases as the day progresses because of solar radiation intensity. The photovoltaic panel efficiency declines at higher temperature. The direct contact heat exchange with RO feed water from photovoltaic panel results in improved performance of photovoltaic cell, at the same time, also tap thermal energy that would have been wasted otherwise. The tapped thermal energy in RO feed water increases its temperature, thus decreases its viscosity and increases the flow performance of reverse osmosis membrane.

Thus the options to improve attractiveness of solar powered reverse osmosis are as shown below:

1. Capitalization of waste thermal energy to elevate the temperature of feed water. The cooling of solar photovoltaic panel can improve the overall panel efficiency from about 6% to upto 40%
2. High recovery process: Increasing recovery decreases the overall energy consumption.
3. Making high flux membrane to decrease overall energy consumption.
4. Decrease the footprint of the total process.

References:

1. Zaher Al Suleimani, V.Rajendran Nair (2000) "Desalination by solar-powered reverse osmosis in a remote area of the Sultanate of Oman", *Applied Energy*, 65, (1–4), 367-380.
2. Murray Thomson, David Infield (2005) "Laboratory demonstration of a photovoltaic-powered seawater reverse-osmosis system without batteries", *Desalination,* 183 (1–3), 105-111.

3. Agustín M. Delgado-Torres, Lourdes García-Rodríguez, Vicente J. Romero-Ternero, (2007) "Preliminary design of a solar thermal-powered seawater reverse osmosis system", Desalination 216 (1–3), 292-305.
4. Guanghui Xia, Qingxuan Sun, Xu Cao, Jiangfeng Wang, Yizhao Yu, Laisheng Wang (2014), "Thermodynamic analysis and optimization of a solar-powered transcritical CO2 (carbon dioxide) power cycle for reverse osmosis desalination based on the recovery of cryogenic energy of LNG (liquefied natural gas)", *Energy,*66, 643-653.
5. Esmail M.A. Mokheimer, Ahmet Z. Sahin, Abdullah Al-Sharafi, Ahmad I. Ali (2013), "Modelling and optimization of hybrid wind–solar-powered reverse osmosis water desalination system in Saudi Arabia", *Energy Conversion and Management,* 75, 86-97.
6. M. Khayet, M. Essalhi, C. Armenta-Déu, C. Cojocaru, N. Hilal (2010), "Optimization of solar-powered reverse osmosis desalination pilot plant using response surface methodology", *Desalination*, 261 (3), 31, 284-292.
7. Eyad S. Hrayshat (2008), "Brackish water desalination by a standalone reverse osmosis desalination unit powered by photovoltaic solar energy", *Renewable Energy*, 33 (8), 1784-1790.
8.Murray Thomson, David Infield (2005), "Laboratory demonstration of a photovoltaic-powered seawater reverse-osmosis system without batteries", *Desalination,*183 (1–3), 105-111.
9. Rashid A. Khaydarov, Renat R. Khaydarov (2007), "Solar powered direct osmosis desalination", *Desalination* 217 (1–3), 225-232.
10. Maedeh P. Shahabi, Adam McHugh, Martin Anda, Goen Ho (2014), "Environmental life cycle assessment of seawater reverse osmosis desalination plant powered by renewable energy", *Renewable Energy,* 67, 53-58.
11.Moomaw, W., P. Burgherr, G. Heath, M. Lenzen, J. Nyboer, A. Verbruggen (2011), " Annex II: Methodology. In IPCC: Special Report on Renewable Energy Sources and Climate Change Mitigation", 10.
12. http://www.globalwaterintel.com/desalination-industry-enjoys-growth-spurt-scarcity-starts-bite/
13. Mingheng Li (2011), "Reducing specific energy consumption in Reverse Osmosis water desalination: An analysis from the first principles" *Desalination* 276, 128–135.
14. Annett Hertel, Ernst Steudle (1997), "The function of water channels in Chara: The temperature dependence of water and solute flows provides evidence for composite membrane transport and for a slippage of small organic solutes across water channels Planta" 202, 324-335.
15. Hiren D. Raval, Subarna Maiti, Ashish Mittal (2014), "Computational fluid dynamics analysis and experimental validation of improvement in overall energy efficiency of a solar photovoltaic panel by thermal energy recovery" Journal of renewable and sustainable energy, 6(3), 033138.

394219 Reverse Electrodialysis: Sustainable Energy from Hydraulic Fracturing Water Recycle

Abstract:

Due to the growing limitations in dependence on fossil fuels, attention has been turned towards searching for alternative means for harvesting energy. One emerging source is energy produced from natural salinity-gradients between freshwater and seawater. Two different processes used to accomplish this are pressure-retarded osmosis and reverse electrodialysis. Significant amounts of research are currently under way regarding pressure-retarded osmosis, however very few studies have been conducted on reverse electrodialysis due to heavy resistance in the diluate chamber of a reverse electrodialysis cell. The focus of this research is to reduce this resistance by using wafers and ion-exchange beads in order to maximize net power gain. Surrogate brackish/saltwater feed streams were used to measure overall power density.

By reducing resistance to levels previously deemed impossible, we are exploring the fundamental limits of separation technology. This research proposes an innovative resolution to high resistance levels through the use of ion exchange wafers to promote ionic movement. Here we show that such wafers have a great impact on the reverse electrodialysis process, and succeeded in achieving a sizeable gain in power density.

Traditional applications include harnessing energy from areas where freshwater bodies meet saltwater, as in an estuary. The future of reverse electrodialysis, however, lies in fracking. Produced water from a fracking well site in Oklahoma was used for power density calculations under realistic conditions. Preliminary results show that our system is capable of producing power densities of up to 10W/m^2, which are five times higher than the current state-of-the-art. Environmental concerns will be all but eliminated if fracking can evolve to be a sort of cyclic process, where brackish fracking water is converted to energy to, in turn, power the overall operation. This poster presentation will model the power increase of electrode ionization and comment on future potential for improvement, especially in the realm of hydraulic fracturing applications.

394215 Modeling of the Aerobic Cometabolic Transformation of Chlorinated Ethenes By the Mycobacterium Elw-1

Abstract:

ELW1 is a recently isolated aerobic bacterium, in the mycobacterium family, that uses isobutene as a growth substrate. ELW1 was also discovered to have the ability to cometabolically transform substances with chemical structures similar to isobutene, such as chlorinated ethenes, into chloride ion and water. Chlorinated ethenes are notorious carcinogens, and groundwater pollutants, whose remediation has been a worldwide problem for over 40 years. Bioremediation of chlorinated ethenes by the augmentation of contaminated soils with microorganisms such as ELW1 is a proposed method of reducing contaminant concentrations in groundwater. To determine the viability of using ELW1 for groundwater remediation, an initial kinetic analysis of transformation rates must be performed. The focus of this research was to determine the rates of contaminant transformation when this organism was exposed to high concentrations of chlorinated ethenes such as trichloroethylene (TCE), *cis*-dichloroethene (c-DCE), and vinyl chloride (VC).

The ELW1 culture was grown and sustained in both batch culture and steady-state chemostat environments with isobutene as a growth substrate on a minimal growth media. Cells that were used for experiments were concentrated by centrifugation, washed with excess media and placed in a new media solution containing high concentrations of the compound to be tested. Initial rates of transformation were determined using gas chromatography to analyze 100μL samples of headspace gas over a period of one to three days to determine the rate of compound loss. The rates at which ELW1 can utilize isobutene were compared with rates of chlorinated ethene transformation, which overall quantifies the practicality of using this organism bioremediation purposes in the field.

Kinetic parameters of these cometabolic processes were determined using an analytical equation based on a modified Monod equation in order to account for decrease in cell activity over time. Results of this investigation indicate that the ELW-1 culture would be effective for the bioremediation of c-DCE and VC, but would be less effective for treated TCE.

394214 The Enzymatic Hydrolysis of Alfalfa Stalks for Use As a Biofuel Resource

Abstract:

Most biofuel resources, especially plant sources, have two key issues: farming space and slow growth. After overexpressing the sucrose phosphate synthase gene (SPS) in Medicago sativa (alfalfa) plants, large differences in phenotype and growth rate were seen in the plant. While the leaves of the alfalfa plants can be used for animal forage, the stems of the alfalfa can be fractionated into sugars that can then be fermented into bioethanols. This technique could solve the growth problem faced by bioethanol producers.

The purpose of the project was to fractionate and analyze the extractability of fermentable sugars from a control and several transgenic species that show overexpression in the SPS gene. Using a strong acid pre-treatment and enzymatic hydrolysis to break down the cellulose into glucose, the alfalfa stalks were fractionated then analyzed using ultraviolet spectroscopy with anthrone as an indicator for glucose content. Two of the three tested transgenic species were shown to have more glucose content per gram than the control species (2-4% more glucose/gram). A further development of the overexpression technique can lead to faster growth, increased concentration of fermentable carbohydrates, and larger crop yields.

UNIVERSITY OF SOUTH CAROLINA
College of Engineering and Computing

SAGE Power through Engineering

Synthesis and Characterization of FeAlCr via High Throughput Techniques

Randy Fang
Jonathan Kenneth Bunn and Dr. Jason Hattrick-Simpers
Department of Chemical Engineering
University of South Carolina
AICHE National Conference 2014
11/17/2014

UNIVERSITY OF SOUTH CAROLINA Honors College

1

Fukushima Daiichi Nuclear Disaster

https://coto2.files.wordpress.com/2011/03/2-fuel-pellet-assembly.jpg

http://diggingintheclay.files.wordpress.com/2011/03/5525887859_e1934af238_o_d1.jpg

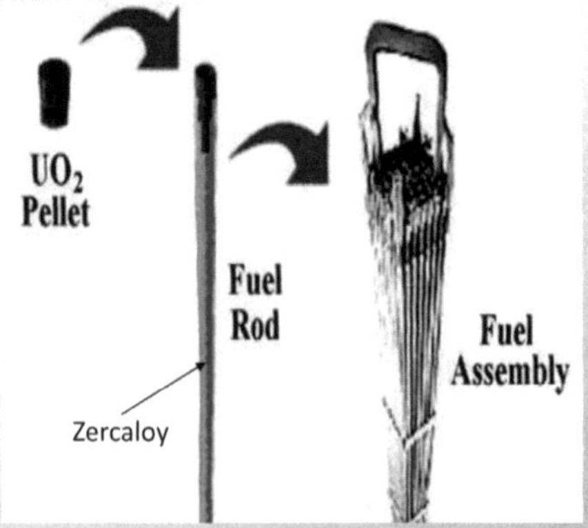

2

FeAlCr Alloy

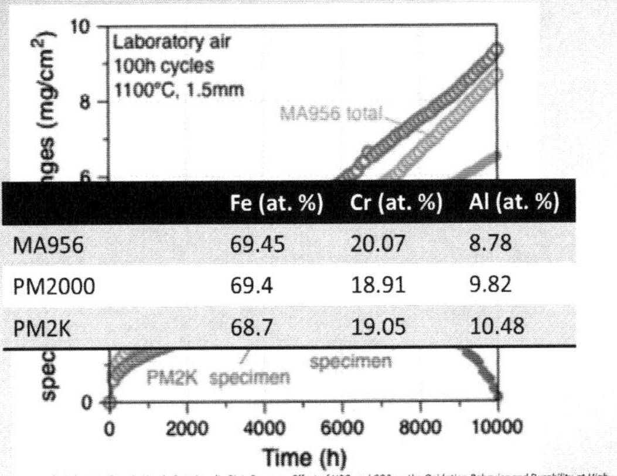

	Fe (at. %)	Cr (at. %)	Al (at. %)
MA956	69.45	20.07	8.78
PM2000	69.4	18.91	9.82
PM2K	68.7	19.05	10.48

Dryepondt, Sebastien Rouaix-Vande Put, Aurelie Pint, Bruce a., *Effect of H2O and CO2 on the Oxidation Behavior and Durability at High Temperature of ODS-FeCrAl*, Oxidation of Metals, Vol 79,2003

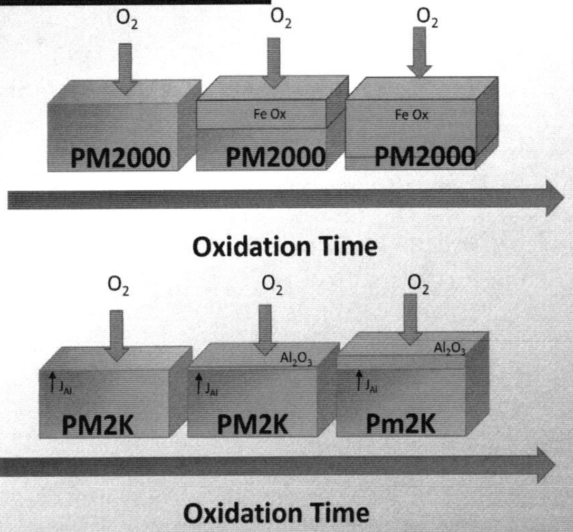

Introduction to High Throughput Techniques

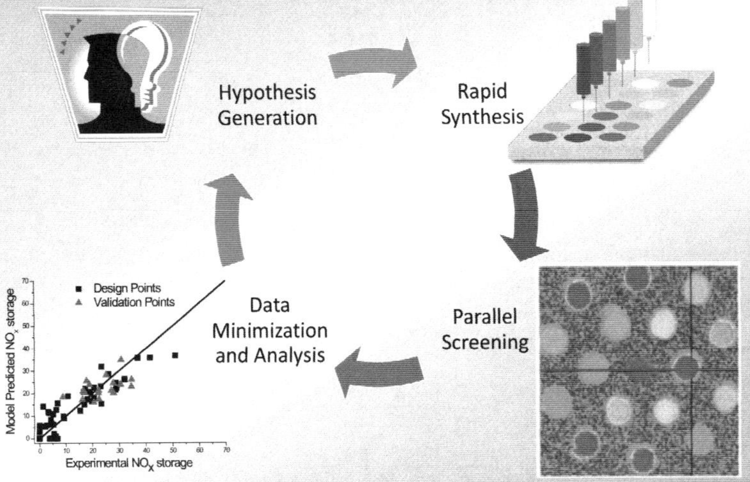

Research Objective

- **Implementing established high-throughput methods for FeAlCr high temperature oxidation resistance screening**
- Non-destructive characterization
- Gain comprehensive understanding of:
  - Time dynamics of oxide formation at high temperature
  - Role of metallurgical phase in promoting growth and continued growth of passivating oxide
- Identification of promising compositions for passivating oxide growth

High Throughput Approach to Explore the Ternary Phase Diagram

Qu Materials Science and Engineering A Vol 562 (2013)

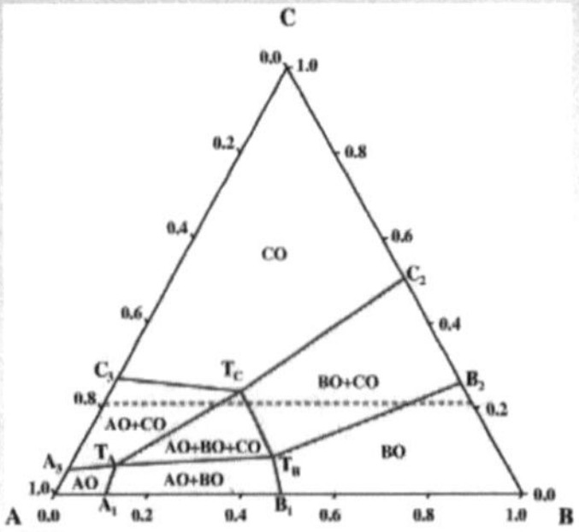

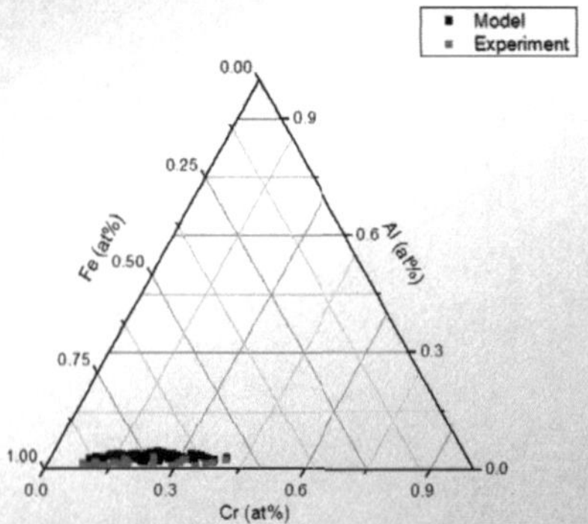

J. K. Bunn, C. J. Metting, and J. Hattrick-Simpers, "A Semi-Empirical Model for Tilted-Gun Planar Magnetron Sputtering Accounting for Chimney Shadowing ", JOM, Invited paper, Accepted January 2014 Special Issue

Sputtering Geometry and Sample Locations

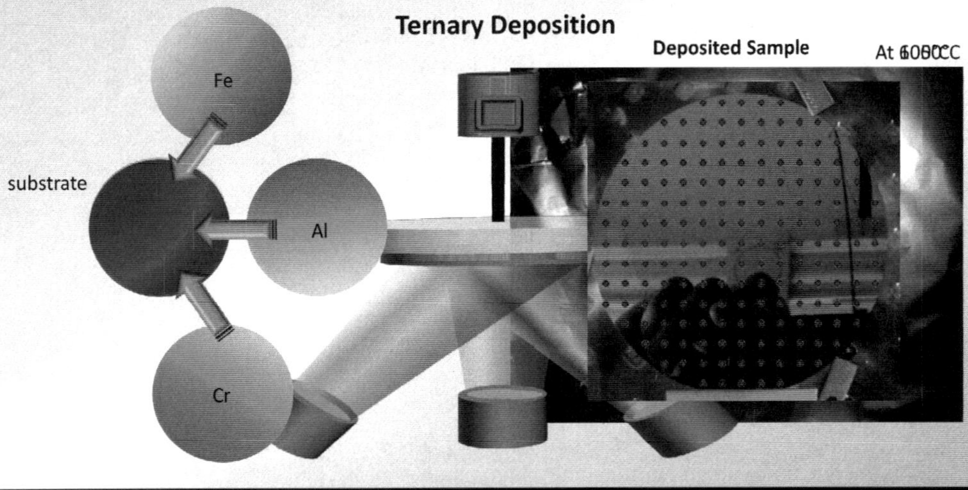

7

Magnetron Sputtering

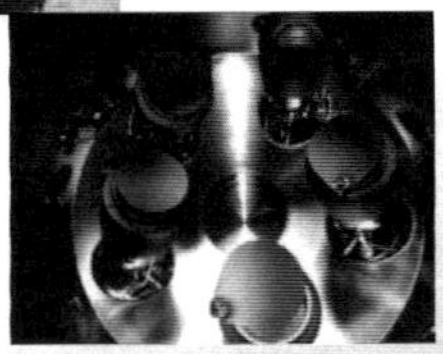

- **5 Guns (only used 3)**
- **DC/RF power sources**
- **Ultra high vacuum**
- **Deposition temperature at room temperature**

8

Experimental Design for High Temperature

Sample fabrication
-12 FeAlCr samples prepared on Al_2O_3 substrate and Ti with adhesion layer
-1 micron nominal thickness

Heat treatment
- Tube furnace treated
- Treated at 1050°C
- Air quenched

After deposition

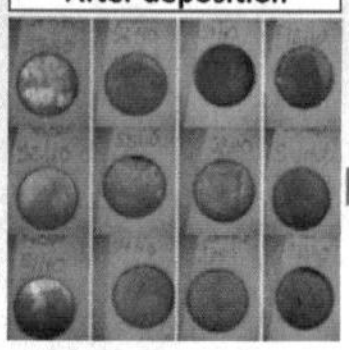

Cycled 7 times for cumulative heat treatments at 1050°C for:
-5 min -10 min
-20 min -40 min
-60 min -120 min
-240 min -360 min

After heat treatments

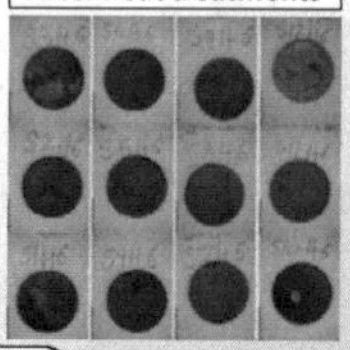

Characterization
-SEM/EDS
-XRD

Characterization Techniques
- **Raman**: Fe_2O_3, Fe_3O_4, and Al_2O_3
- **XRD**: FeCrAl, Surface Oxides, and the substrate
-**Optical Microscopy**

Oxide Formation at 5 minutes, 1050°C

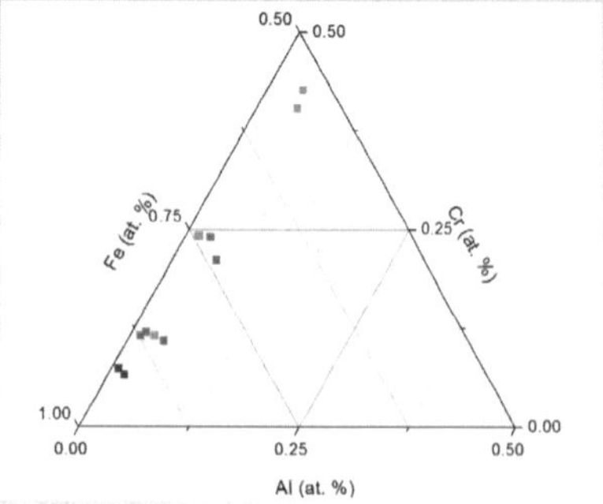

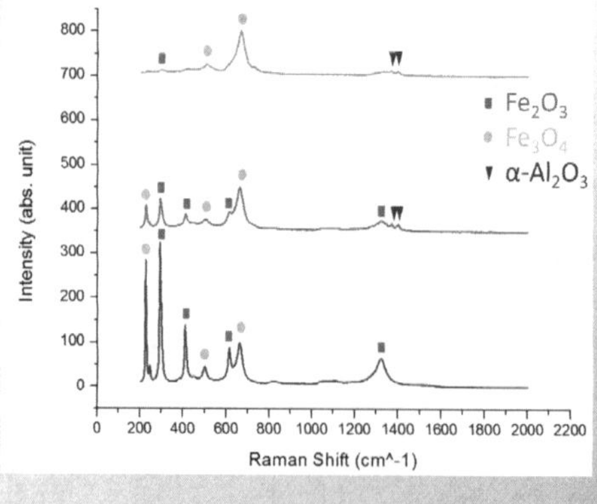

Blue: mostly Fe_2O_3, some Fe_3O_4
Red: mostly Fe_3O_4, some Fe_2O_3, some α-Al_2O_3
Green: very little Fe_2O_3, some α-Al_2O_3, some Fe_3O_4

- No α-Al_2O_3 growth was observed when the composition is Fe heavy

Oxide Formation at 360 minutes, 1050°C

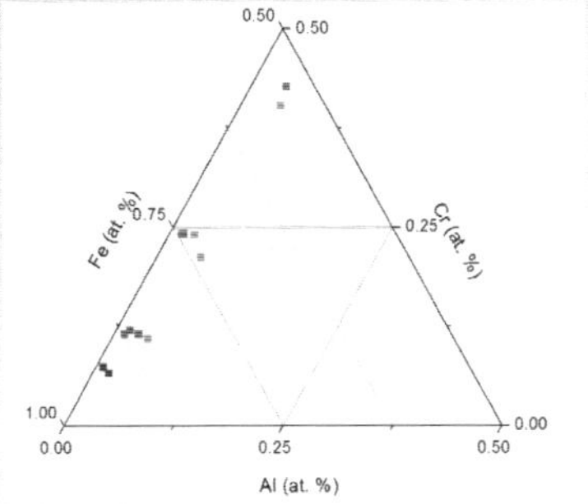

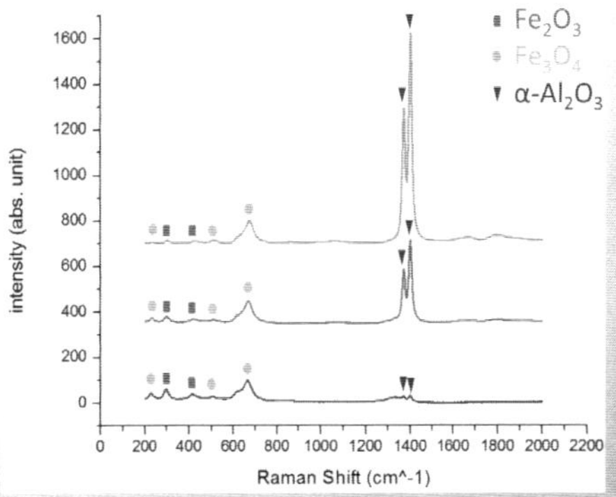

Blue: some Fe_2O_3 and Fe_3O_4, very little α-Al_2O_3
Red: some α-Al_2O_3, some Fe_3O_4
Green: majority α-Al_2O_3, some Fe_3O_4

- Large amount of α-Al_2O_3 growth was observed at the Cr heavy region, indicating that Cr promotes the growth of the alumina oxides

Experimental Design at Low Temperature

Sample fabrication
-FeAlCr samples prepared on Si substrate with Ti diffusion barrier
-1 micron nominal thickness

Cycled 6 times for cumulative heat treatments 1 hour each (with the exception of 600°C):

-room temperature	-200°C
-300°C	-400°C
-500°C	-600°C (for 2hr)

Characterization
-*in-situ* GIXRD

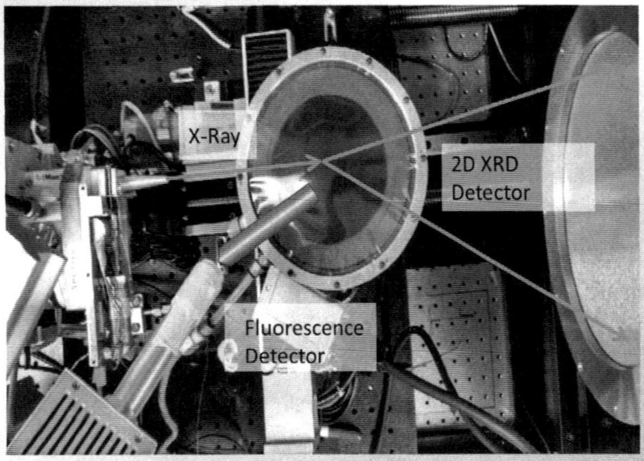

Oxide Growth Study through *in-situ* XRD

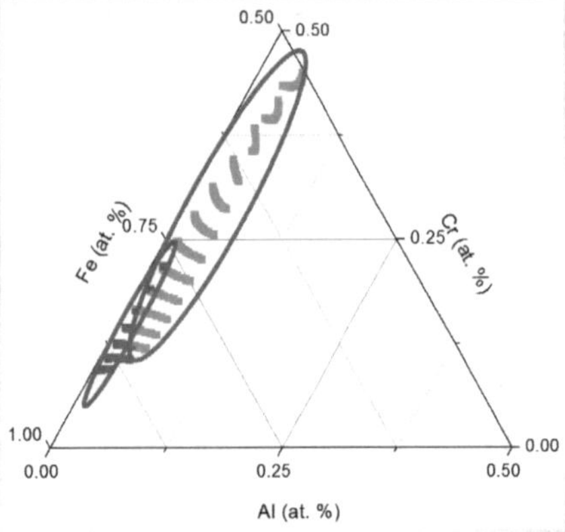

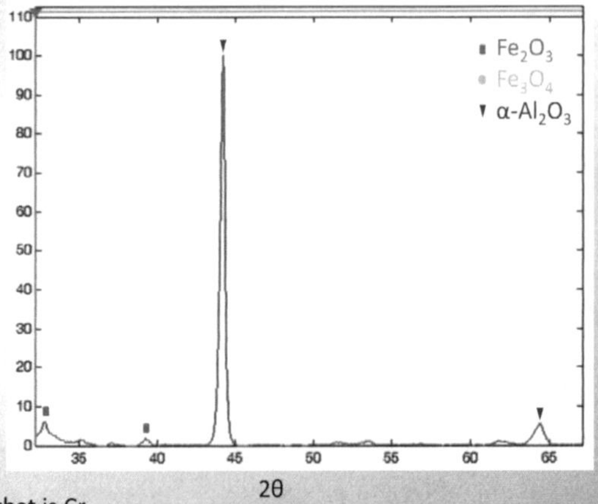

- Iron oxides are suppressed when the composition that is Cr rich (small amount are observed)

Comparison of Trends Between Low and High Temp Oxidation

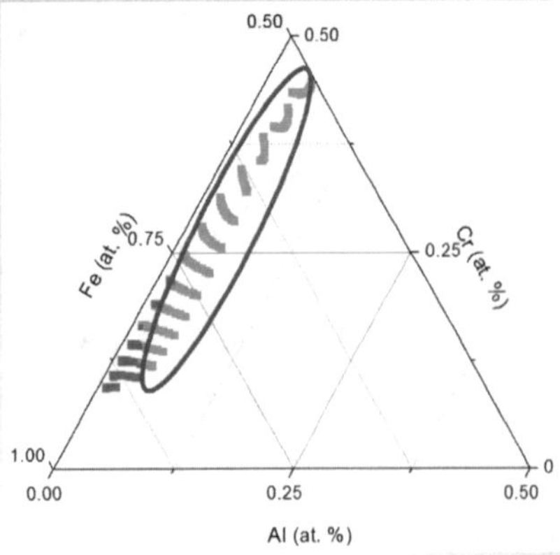

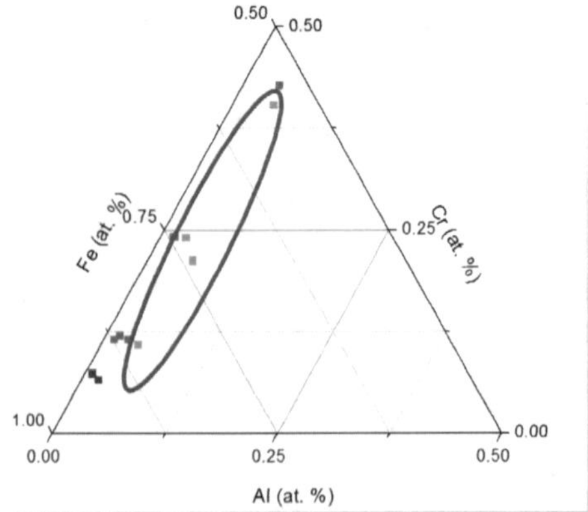

- XRD and Raman matched up where the more Al and Cr rich region are a lot easier to form α-Al_2O_3

Summary

- Probed the temporal effect of temperature on the metallurgic and oxide phase transitions
- A method for the **initial high-throughput and non-destructive screening** of FeCrAl thin film was developed
- **Cr promotes** α-Al_2O_3 growth on the surface of the samples to form a protective layer preventing further development of other oxides
  - This indicates that Cr rich FeCrAl are more favored to operate under high temperature conditions
- Oxidation studies indicate that severe oxide growth was not observed until 400°C
- The methodology can be applied to more complicated systems for high temperature oxidation resistant materials.
- Detailed studies of **oxidation and phase transitions of single compositions at different temperatures** will refine the results

15

Acknowledgement

- **Dr. Jason Hattrick-Simpers**
- **Jonathan Kenneth Bunn**
- **SmartState Center for Strategic Approaches to the Generation of Electricity**
- **USC Honors College (Science Undergraduate Research Fellowship Program)**
- **University of South Carolina**
  - **College of Engineering and Computing Science**

This project was funded in part by the South Carolina Honors College

16

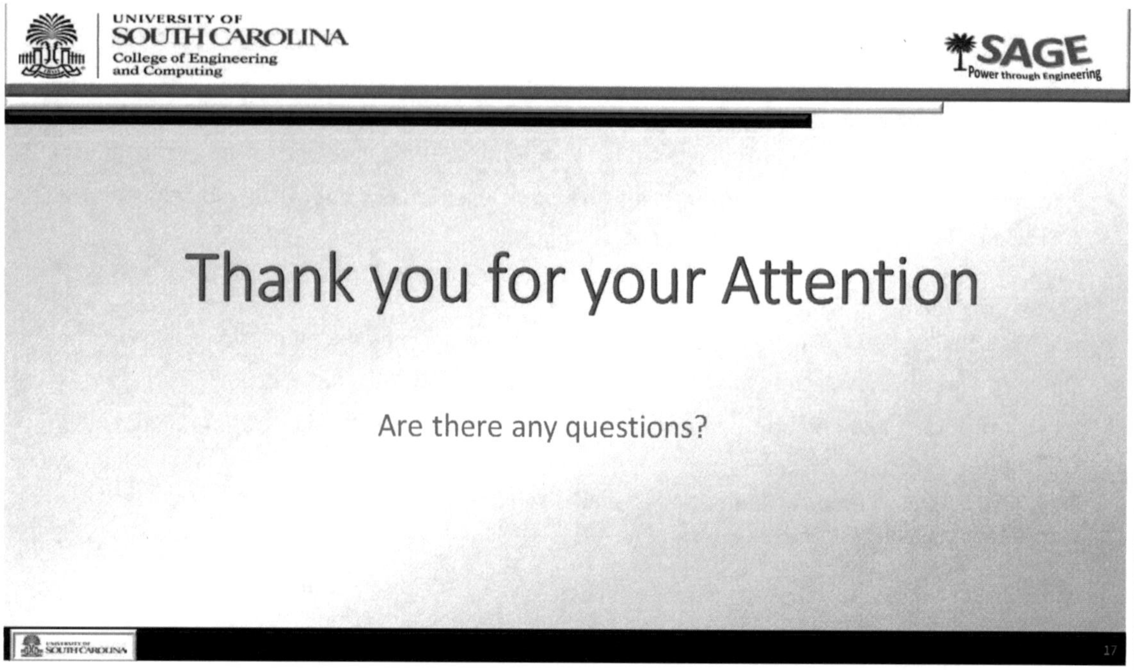
UNIVERSITY OF
SOUTH CAROLINA
College of Engineering
and Computing
SAGE
Power through Engineering
Thank you for your Attention
Are there any questions?
17

Crossing Borders and Communities for Healthy Water

Joshua Gomez, Meng Zhou, Dr. Shuguang Deng

Chemical and Materials Engineering Department, New Mexico State University, Las Cruces, New Mexico, 88003

In Partnership with:

NM Water Resources Research Institute

NMSU Chemical Engineering Department

The New Mexico Alliance for Minority Participation

New Mexico Environment Department

Border Partners (Non-profit)

EPA 2020

National Science Foundation

New Mexico State University NM STATE

Puerto Palomas, Chihuahua, MX

- Located just south of Columbus, NM.
- Population of 4,688.
- Arsenic contamination (0.030-0.1 mg/L).
  - Horrible Carcinogen.
- Fluoride contamination (5-40 mg/L).
  - Bone Fluorosis and neurological damage.
- Concentration depends entirely on well water.
  - Naturally occurring concentrations of arsenic and fluoride exceed the permissible limits of Mexico and the U.S.

New Mexico State University NM STATE

New Mexico Transboundary Aquifers

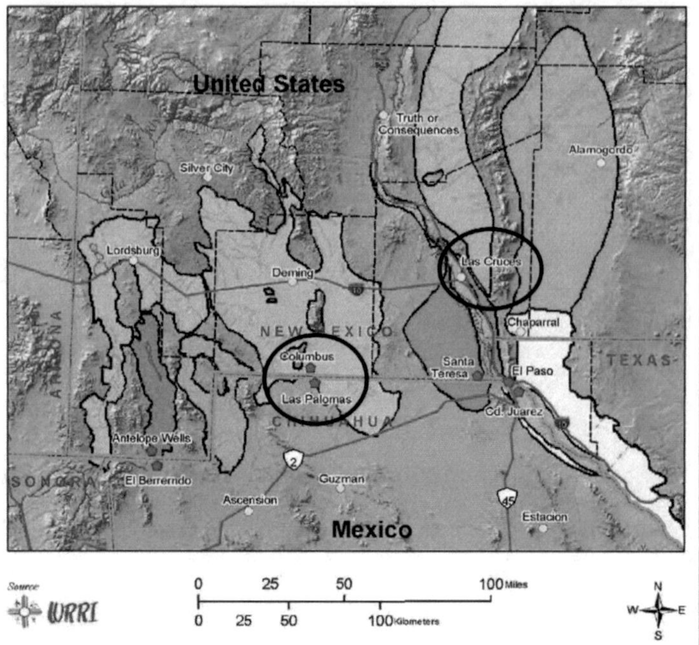

New Mexico State University

Proposed Solution

- Homemade water filter
  - Low cost <$15.00
  - Easy to assemble
  - Removes arsenic
  - Removes fluoride
  - Utilizing activated carbon (AC)
  - Utilizing activated alumina(AA)
    - Iron coated
  - Improves the taste

New Mexico State University

Filter Designs

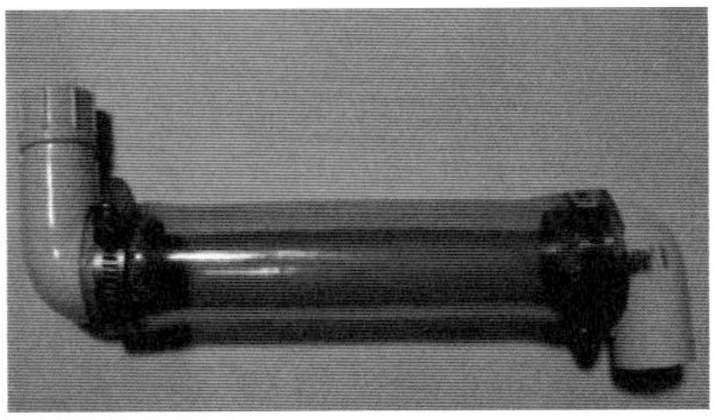

- 1 filter of the final design costs $5.00 to make.
- The Activated Alumina costs $1/kg.
- The Activated carbon used cost $35/lb.

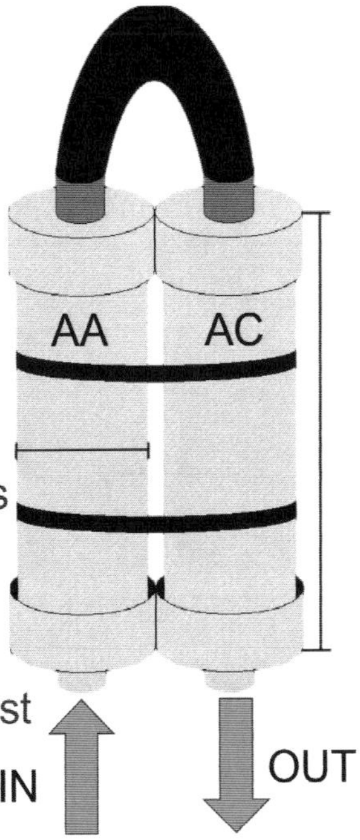

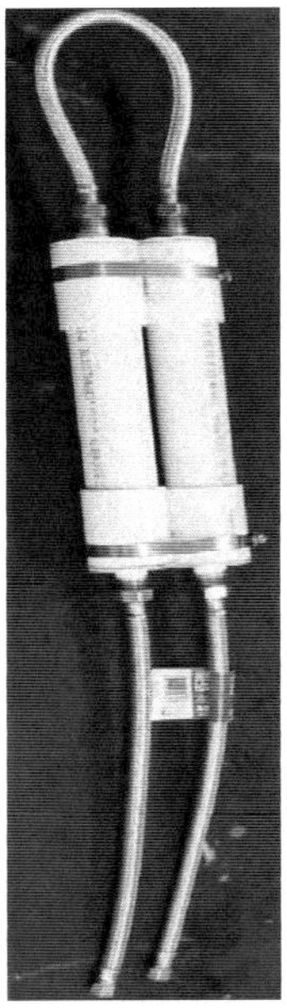

New Mexico State University

Our Approach

NMSU demonstrated to Palomas science teachers how to assemble a simple water filter.

Science teachers Juan Carlos Huerta and Joel Carreon pose with the filter they assembled for installation at the Palomas primary school.

New Mexico State University

What's Working

Hall Environmental Laboratory, Albuquerque

CONTAMINANT	RESULTS	STANDARDS					
		U.S.	Exceeded	%	Mexico	Exceeded	%
Fluoride							
Unfiltered	4.90 mg/L	4.00 mg/L	Y	22.5	1.50 mg/L	Y	226.7
Filtered	0.240 mg/L	4.00 mg/L	N	-	1.50 mg/L	N	-
Arsenic							
Unfiltered	0.030 mg/L	0.01 mg/L	Y	200.0	0.02 mg/L	Y	20.0
Filtered	ND	0.01 mg/L	N	-	0.02 mg/L	N	-

ND = Not Detected

New Mexico State University NM STATE

Optimization of Algae Derived Activated Carbon (AC)

- Low cost alternative to commercial activated carbon.
- More effective adsorbent than commercial carbon.
- Chlorella algae is used as a carbon source.
- Potassium Hydroxide and Iron (II) Chloride
  - Activating agent.
- Carbonization times of 500° C, 600° C, and 700° C.
- Shaker experiments to determine most effective adsorbent for Arsenic(III).

New Mexico State University NM STATE

Carbon Experimental data

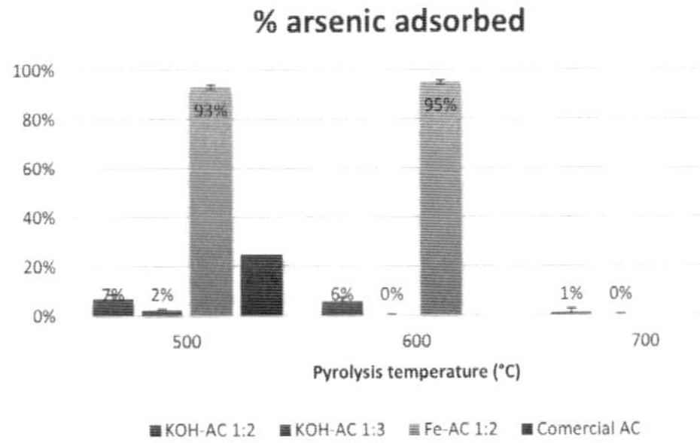

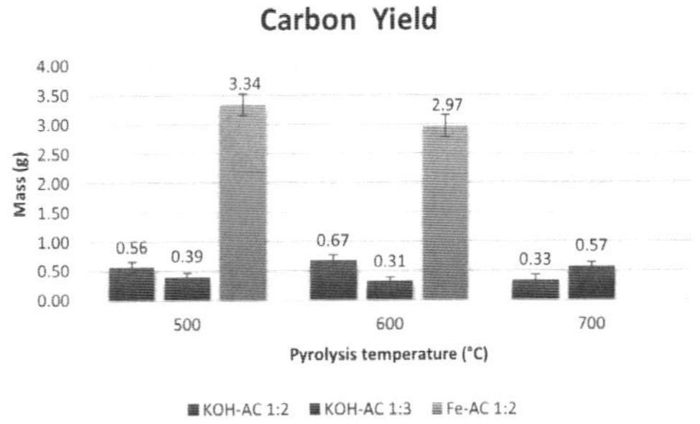

adsorption capacity (mg/g)			
	500°C	600°C	700°C
KOH-AC 1:2	0.082	0.067	0.010
KOH-AC 1:3	0.024	0.000	0.000
Fe-AC 1:2	1.143	1.168	-
Comercial AC	0.275	-	-

$$capacity = \frac{(InitialConcentration - FinalConcentration)Volume}{Mass\ of\ adsorbent}$$

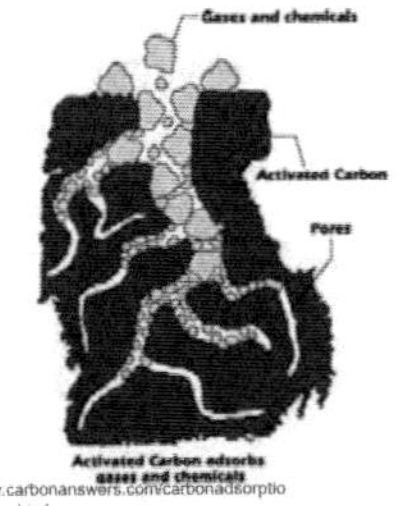

Fasci, M.
http://www.carbonanswers.com/carbonadsorptionabsorption.html

New Mexico State University

X-Ray Difraction Pattern

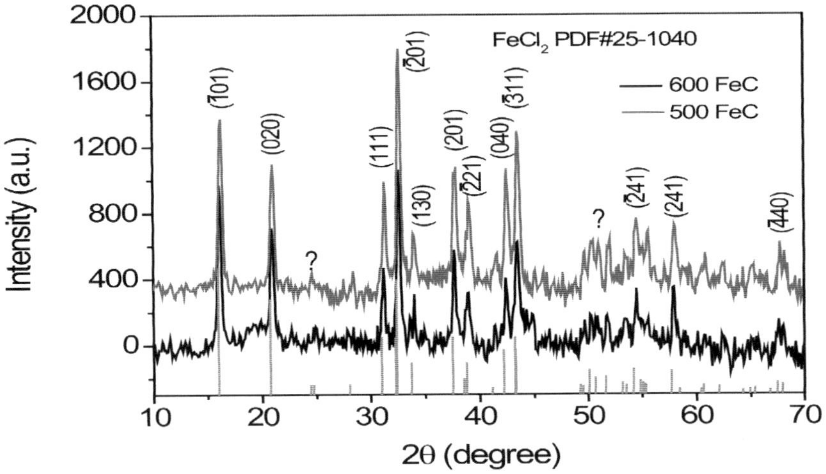

New Mexico State University

Scanning Electron Microscope Imaging

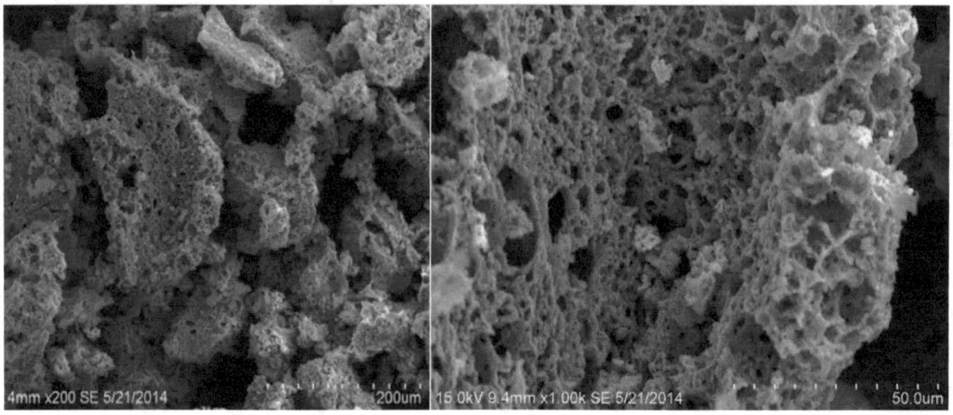

New Mexico State University

Unwashed FeCl(II) Activated Carbon Textural Properties

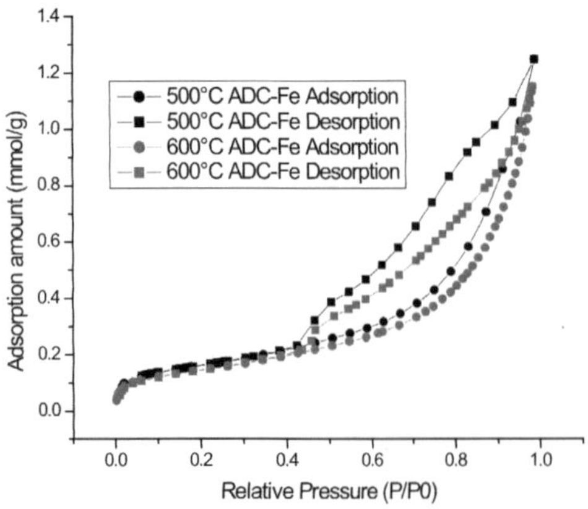

N_2 adsorption and desorption isothermal @ 77K for unwashed samples.

	500 ° C FeCl2/ADC	600 ° C FeCl2/ADC
BET surface area (m^2/g)	12.9893	11.6336
Average pore size (nm)	8.6408	6.5769
Pore volume (cm^3/g)	0.0394	0.0476

New Mexico State University

Washed FeCl(II) Activated Carbon Textural Properties

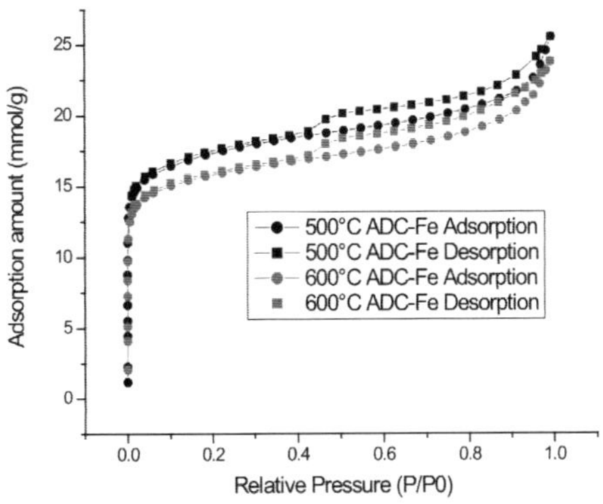

	500 ° C ADC (after washing)	600 ° C ADC (after washing
BET surface area (m^2/g)	1326.0368	1208.1831
Average pore size (nm)	7.8704	7.9340
Pore volume (cm^3/g)	0.3362	0.3305

N2 adsorption and desorption isothermal @ 77K for washed samples.

New Mexico State University

Shaker Experiments

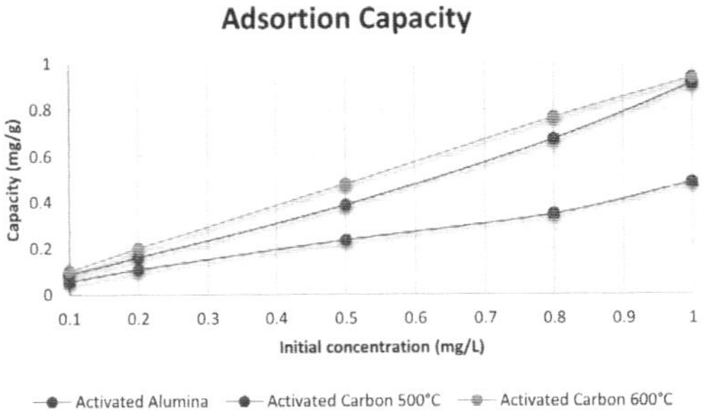

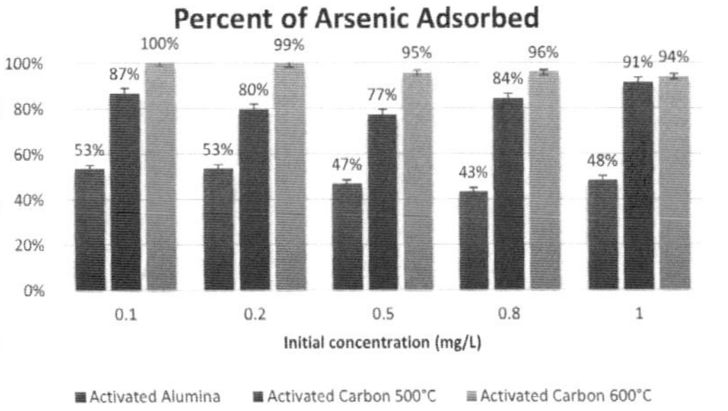

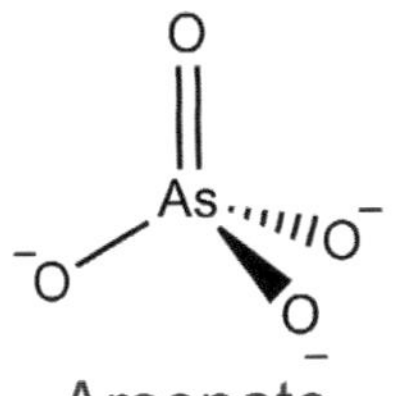

Arsenate

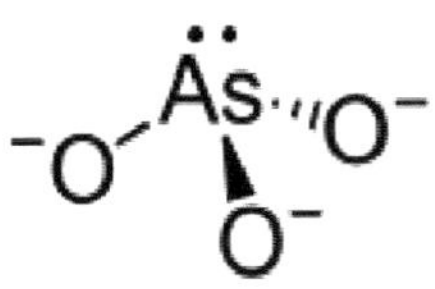

Arsenite

New Mexico State University

Conclusions

- Filters are still being made in Palomas by the teachers today and will continue to be made.
- The results of the shaker experiments and textural property data shows that the carbon activated with Iron (II) Chloride is a viable replacement.

New Mexico State University

Next Steps

- Further Development of the filter.
- Reach out to other communities.
- Advance research with optimized carbon.

New Mexico State University

Acknowledgments

- Dr. Shuguang Deng, Chemical & Materials Engineering dept.
- Meng Zhou, Chemical & Materials Engineering dept.
- Dr. David A. Rockstraw, Chemical & Materials Engineering dept.
- Erin Ward, Water Resources Research Institute.
- Mark Chidester, Chemical & Materials Engineering dept.
- Peter Edmunds, Border Partners.

NSF HRD # 1305011

Deng's Renewable Energy and Advanced Materials

New Mexico State University

Questions?

Email: jgomez38@nmsu.edu

New Mexico State University

394186 Ferroelectric BTO on Si (001) for High-Efficiency Solar Cell Heterostructures

Abstract:

Ferroelectric oxides such as barium titanate (BTO) have recently become a topic of interest in relation to thin-film solar cell production due to the material's stability at a wide range of temperatures, its spontaneous polarization, and the potentially low-cost of production. Integration of BTO on silicon could enable the creation of highly efficient solar cells, as BTO's ferroelectric properties encourage the detachment of photo excited electrons. We have previously demonstrated successful integration of BTO on silicon carbide (SiC) substrates, using a magnesium oxide (MgO) heteroepitaxial layer to enable single-crystal orientation of the BTO film deposited on SiC.[1] BTO has also been successfully grown on silicon substrates, but with the use of strontium titanate (STO) and MgO buffer layers to resist formation of SiO_2.[2] Integration of BTO on Si with a single MgO interface layer would decrease interfacial losses in a final solar cell device. Based on previous work with SiC, we believe a single interface can be possible by controlling the surface interactions during film growth.

In this research, MgO is grown on Si (001) substrates in an ultrahigh vacuum (UHV) environment using a number of growth methods including molecular beam epitaxy (MBE) and sputtering. In comparing differences in surface structure using reflection high-energy electron diffraction (RHEED) and surface chemistry using X-ray photoelectron spectroscopy (XPS), we can understand surface interactions between Si, Mg, and O. By understanding the relationship between atomic level interactions and ultimate film characteristics, we can engineer the most effective process for BTO integration on Si. The substrates are cleaned using wet chemicals to create a hydrogen-terminated surface that resists contamination from laboratory exposure. XPS and RHEED are used to verify the lack of contamination on the surface and proper crystallographic orientation. Contaminant levels of 7.61 % oxygen (0.15 O/Si ratio) and 18.0% carbon (0.31 C/Si ratio) are the most successful results that have thus far been attained. XPS and RHEED will serve to confirm proper growth of MgO (001) on Si and determine thickness of the MgO film. BTO is then grown on the correctly-oriented MgO film using MBE, as has already been demonstrated on MgO/SiC substrates.[1] This presentation will discuss cleaning and growth conditions for successful production of high quality BTO on Si using the single MgO interface layer.

393873 Natural Gas, the Bridge Fuel

Abstract:

Medium and heavy duty trucks and buses produce almost 18 percent of all carbon emissions from transportation. Alternative energy sources need to be used on a larger scale to reduce these emissions, as well as to reduce dependence on foreign oil. Natural gas is the future of fuel in large vehicles. This paper tackles different issues surrounding the usage of natural gas as the main fuel for natural such as public awareness, public policy, existing infrastructure, consumer choice and controversies surrounding hydraulic fracturing. Current pilot programs, in which public city buses are using natural gas as the fuel of their choice, will be analyzed to determine the best way to implement these programs in other areas of the United States as well.

397811 The Federal Role in Fostering an Innovative U.S. Energy Ecosystem

Abstract:

Although renewable sources of energy – solar, wind, hydro, and geothermal – have been around for over one hundred years, they only generate 12.5 percent of electricity in the United States today. Renewable technology has advanced in cost, reliability, performance, and efficiency, but not enough to allow renewables to be a large contributor to electricity generation and widely used in the U.S. Renewables need continued innovation and progression to become more efficient, more reliable, and more cost competitive with fossil fuels, as well as expanding the possibilities of their use. Increased R&D funding for renewable sources is the key to advancing the technology and giving renewables the tools it needs to provide a larger portion of electricity in the U.S.

With renewables generating a larger portion of electricity, the U.S. can reach environmental sustainability, energy security, and a new economic opportunity. In reference to reaching environmental sustainability, renewables do not emit carbon and greenhouse gases, which is important for a cleaner future and mitigating climate change. Because the "fuel" for renewables includes the sun and wind, electricity generation from these sources does not depend on obtaining fossil fuels inside or outside of our country. They also extend the lifetime of these resources to be used for base load power, providing energy security to the U.S. Finally, if the U.S. is able to quickly and effectively develop competitive renewable technologies and deploy them into the electricity sector, it can provide competitive renewable technologies to the global community, boosting the economy and maintaining world leadership.

Although renewables have the capability to provide many benefits for the U.S., they face obstacles in receiving the funding they need to be widely used. These obstacles include climate change skeptics, the fossil fuel industry, the polarized political environment, and having to cut funding for other areas such as education and defense in order to allocate more money to R&D. The demand for renewables has also been relatively weak because electricity generated by renewables is no different than electricity generated by fossil fuels when the consumer uses it. The main way for renewables to be competitive with fossil fuels is through the cost and since wind is the only renewable that is cost competitive with fossil fuels, a strong customer demand has not been present to spur large private sector investment in renewable technologies.

In order for renewables to accomplish necessary advancements and contribute a larger portion of electricity to the U.S., financing to R&D for the technologies needs to be increased through government funding, capital generated through a carbon tax, public-private partnerships, and incentives for investment in the technology and its R&D. A commitment to innovating renewables in order to reach environmental sustainability, energy security, and a new economic opportunity is necessary from governmental officials to make the tough decisions advocating for widespread use of renewables in the U.S.

397856 A Biowall for Improving Indoor Air Quality

Abstract:

The Biowall is a living plant wall that filters air in the HVAC system of a home or small commercial building. This new technology improves indoor air quality be removing Volatile Organic Compounds and has the potential for energy savings by reducing the need for heating and cooling large amounts of outside air in a building. The presentation provides an overview of biofiltration for buildings and discusses the new research being conducted in this emerging field.

Ionic liquid electrolytes have been proposed as a means of improving the safety and cycling behavior of advanced lithium batteries; however, the properties of these electrolytes under high lithium doping are poorly understood. Here, we employ both polarizable molecular dynamics simulation and experiment to investigate the structure, thermodynamics and transport of three potential electrolytes, N-methyl-N-butylpyrrolidinium bis(trifluoromethylsufonyl)imide ([pyr14][TFSI]), N- methyl-N-propylpyrrolidinium bis(fluorosufonyl)imide ([pyr13][FSI]), and 1-ethyl-3--methylimidazolium boron tetrafluoride ([EMIM][BF4]), as a function of Li-salt concentration and temperature. Structurally, Li^+ is shown to be solvated by three anion neighbors in [pyr14][TFSI] and four anion neighbors in both [pyr13][FSI] and [EMIM][BF4], and at all levels of x_{Li} we find the presence of lithium aggregates. Furthermore, the computed density, diffusion, viscosity, and ionic conductivity show excellent agreement with experimental data. While the diffusion and viscosity exhibit a systematic decrease and increase, respectively, with increasing x_{Li}, the contribution of Li^+ to ionic conductivity increases until reaching a saturation doping level of $x_{Li} \sim 0.10$. Comparatively, the Li^+ conductivity of [pyr14][TFSI] is an order of magnitude lower than that of the other liquids, which range between 0.1-0.3 mS/cm. The differences in Li^+ transport are reflected in the residence times of Li^+ with the anions, which are revealed to be much larger for [pyr14][TFSI] (up to 100 ns at the highest doping levels) than in either [EMIM][BF4] or [pyr13][FSI]. Finally, we comment on the relative kinetics of Li^+ transport in each liquid and we present strong evidence for transport through anion exchange (hopping) as opposed to the net motion of Li^+ with its solvation shell (vehicular).

378085 Toward Understanding the Structural Heterogeneity and Ion Pair Stability in Dicationic Ionic Liquids

Abstract:

The structural and dynamical properties of dicationic ionic liquids (DILs) 1,3- bis[3- methylimidazolium-1-yl]alkane bis(trifluoromethylsulfonyl)amide $[C_n(mim)_2](Tf_2N)_2$ were investigated by molecular dynamics (MD) simulation in combination with small-angle X-ray scattering (SAXS) measurement. Enhanced spatial heterogeneity is observed with the prolongation of chain length in DILs, accompanied with an increased heterogeneity order parameter (HOP). The structure function calculated from MD simulation agrees qualitatively with that obtained from SAXS. Temperature variation results in slight influences on the local structural assembly of DILs and all the three peaks at 0.4 $Å^{-1}$, 0.9 $Å^{-1}$ and 1.4 $Å^{-1}$ of the structure function shift towards low Q as temperature increases. The change in peak position by the temperature is also supported by the temperature-independent HOP values. The prepeak shifts are in opposite directions in DILs and MILs as temperature increases. This is revealed in MD simulation for the first time, possibly indicating an RTIL-specific prepeak shifting. Furthermore, the high ion pair/ion cage stability in DILs is indicative of the high thermal stability and insensitivity of structural heterogeneity to temperature variation, which can likely be attributed to the stronger Columbic network interaction in DILs due to the additional charges in dications.

391459 Development of Human Health Impact Characterization Factors for Ionic Liquids

Abstract:

In this presentation we outline a methodology to derive human health characterization factors for ionic liquids that can be used in life cycle assessment (LCA) methods to study the health impacts of ionic liquids. Current LCA impact assessment methods such as TRACI do not have human health characterization factors for ionic liquids. Therefore, at this point health impacts associated with processes that use and release ionic liquids to the environment cannot be considered within an LCA framework. Further, there is very little LD50 data -which is needed for deriving the characterization factors- available for ionic liquids. In this presentation we outline an approximate method based on correlations between in vitro data and LD50 that can be utilized to derive ionic liquid characterization factors. The method is illustrated with examples.

383457 Binary Phase Behavior of Lipidic Ionic Liquids

Abstract:

Ionic liquids, molten salts which melt below 100°C, are a unique class of compounds that have been the subject of significant interest and development in recent decades. As they are comprised of ions, they typically exhibit low to negligible volatility. Because of this, ionic liquids are of great interest as environmentally benign process solvents which do not generate fugitive emissions. Additionally, since the cations, anions or both are typically comprised of organic ions, the full palette of synthetic organic chemistry can be utilized to tune the chemical and physical properties of these compounds, allowing for customizable solvents, catalysts and, more recently, active pharmaceutical ingredients.

The ionic nature of these species typically results in solvent properties similar to those of polar organic solvents, with high solubilities for polar to moderately polar compounds and low solubilities for non-polar species. Recently, however, there has been interest in developing ionic liquids with significant non-polar content for use as solvents and gas absorbents. To achieve "non-polar-like" solvent properties, the non-polar content of the ions can be increased; this can be accomplished by adding a long alkyl chain to the central ion. However, this often results in significant melting point increases and compounds that are room temperature solids.

Our group has described two types of lipidic ionic liquids that are room temperature liquids; these species are inspired by and derived from natural lipids. The first type incorporates *cis* unsaturations into a C_{16}-C_{20} alkyl chain appended to a central ion.[1] The double bond causes a "kink" in the alkyl chain which frustrates packing and decreases the melting point, analogous to natural triacylglycerols with more unsaturations having lower melting points than triacylglycerols with fewer unsaturations. The second type, designed to be more oxidatively stable, have been synthesized using thiol-ene "click" chemistry and have thioehter linkages in place of the double bonds.[2] Each type has significant non-polar content while maintaining low melting points. Previously, we have measured and correlated the pure component thermophysical properties of several representative examples of these lipidic ionic liquids.[3]

In the current work, we describe the binary liquid/liquid equilibria of these species with a variety of molecular compounds ranging in polarity from methanol to hexane, including several homologous series. Of particular interests are the trends observed with respect to the size and shape of the molecular species. We postulate that the structures that arise in the hetereogeneous nanodomains (segregated polar and non-polar domains that have been described experimentally[4] and through molecular dynamic simulations[5]) play a key role in the overall solubility of the nonpolar compounds. In addition to liquid/liquid equilibrium, vapor/liquid equilibrium of these systems will also be presented.

1. (a) Murray, S. M.; O'Brien, R. A.; Mattson, K. M.; Ceccarelli, C.; Sykora, R. E.; West, K. N.; Davis, J. H., Jr., The Fluid-Mosaic Model, Homeoviscous Adaptation, and Ionic Liquids: Dramatic Lowering of the Melting Point by Side-Chain Unsaturation. *Angew. Chem., Int. Ed.* **2010,** *49* (15), 2755-2758; (b) Mirjafari, A.; Murray, S. M.; O'Brien, R. A.; Stenson, A. C.; West, K. N.; Davis, J. H., Structure-based tuning of Tm in lipid-like ionic liquids. Insights from Tf2N- salts of gene transfection agents. *Chem. Commun.* **2012,** *48* (60), 7522-7524; (c) Mirjafari, A.; O'Brien, R. A.; Murray, S. M.; Mattson, K. M.; Mobarrez, N.; West, K. N.; Davis, J. H., Jr., Lipid-inspired ionic liquids containing long-chain appendages: novel class of biomaterials with attractive properties and applications. *ACS Symp. Ser.* **2012,** *1117* (Ionic Liquids), 199-216; (d) Kwan, M.-L.; Mirjafari, A.; McCabe, J. R.; O'Brien, R. A.; Essi, D. F. I. V.; Baum, L.; West, K. N.; Davis, J. H., Jr., Synthesis and thermophysical properties of ionic liquids: cyclopropyl moieties versus olefins as Tm-reducing elements in lipid-inspired ionic liquids. *Tetrahedron Lett.* **2013,** *54*(1), 12-14.

2. Mirjafari, A.; O'Brien, R. A.; West, K. N.; Davis, J. H., Jr., Synthesis of New Lipid-Inspired Ionic Liquids by Thiol-ene Chemistry: Profound Solvent Effect on Reaction Pathway. *Chem. - Eur. J.* **2014**, Ahead of Print.

3. Murray, S. M.; Zimlich, T. K.; Mirjafari, A.; O'Brien, R. A.; Davis, J. H., Jr.; West, K. N., Thermophysical Properties of Imidazolium-Based Lipidic Ionic Liquids. *J. Chem. Eng. Data* **2013,** *58*(6), 1516-1522.

4. Triolo, A.; Russina, O.; Bleif, H.-J.; Di, C. E., Nanoscale segregation in room temperature ionic liquids. *J Phys Chem B* **2007,** *111*(18), 4641-4.

5. Canongia Lopes, J. N. A.; Padua, A. A. H., Nanostructural Organization in Ionic Liquids. *Journal of Physical Chemistry B* **2006,** *110* (7), 3330-3335.

386955 The Effects of Ionic Liquid Structure on the Thermodynamics of Cellulose Dissolution

Abstract:

A big challenge in the production of lignocellulosic biofuels lies in the selection of an optimal pretreatment solvent. A number of ionic liquids (ILs) have been identified that can dissolve cellulose effectively and many more are theoretically possible. Tailoring these solvents has the potential to increase performance and reduce costs, however without a much deeper understanding of the structure-property relationships within ILs this remains a difficult task. Using molecular dynamics we systematically examine the effect of chemical structure on the free energy of dissolution of crystalline bundles of cellulose. We focus on 15 different ionic liquid combinations and calculate the thermodynamic differences between the crystalline and dissolved states. Anions of 3 different sizes are paired with 5 different imidazolium-based cations with differing tail length. Entropies are calculated using the two-phase thermodynamic (2PT) method, allowing us to identify each of the contributions arising from the anion, the cation and cellulose as well as their translational, rotational and vibrational contributions. In doing so we show how changes to the chemical structure affects the overall energy and enthalpy as well as quantify the role that solvent entropy plays in the dissolution process. The free energy differences obtained from the simulations agree well with experimental solubility measurements and we show how these methods can be used to accurately predict solubilities a priori.

The Z-bond interaction and process innovation based on ionic liquid

Suojiang Zhang*, Xiangping Zhang, Kun Dong

Beijing Key Laboratory of Ionic Liquids Clean Process, State Key Laboratory of Multiphase Complex System, Institute of Process Engineering, Chinese Academy of Sciences, Beijing, China

Email: sjzhang@ipe.ac.cn

Ionic liquids (ILs) have been studied extensively as green solvents, catalysts and electrolytes due to their unique properties. Our recent research discovered that it was mainly the Z-bond interaction in ionic liquids that determine the properties of ILs which are inherently different from the conventional molten salt and aqueous solution. Thus, understanding on the Z-bond features in ionic liquids will be helpful for the design of new ILs with specific structure and function, and then developing novel industrial reaction and separation processes taking ILs as catalysts and absorbent or solvent.

Focusing on the Z-bond and the relationship between IL's structure and property, the following research has been conducted. 1) The geometries, interaction energies, charge transfer and frontier MOs of ion pairs in ILs were calculated by DFT method. Due to the electrostatic coupling, the H-bond in ILs showed the co-valent feature, which we defined as Z-bond and proved using far-IR spectra. 2) Using molecular dynamic method and experimental characteristic, the ionic clusters existed in ILs at nano-level have been proved and the influences of cluster on the physicochemical of pure ILs and IL mixtures have been studied. 3) As important properties for scaling up a process, the gas bubble behavior in IL system, such as bubble size distribution, gas holdup, Sauter diameter were investigated, and then the mass transfer properties were measured and predicted. It was also found that Z-bond action and nano ionic clusters make a great contribution to the mass transfer properties in ionic liquid system. 4) The methodology of IL-based process design from molecule to system was proposed and applied to develop several novel technologies, such as cleaner process of Methyl methacrylate (MMA), production of ethylene glycol, CO_2 capture, flue gas desulfuration and NH_3 recovery, were developed in our group, then the main progresses were discussed in this work.

Keywords: Hydrogen bond, ionic liquids, ionic cluster, gas separation, catalysts

References

Zhang S, et al. A new class of ion-ion interaction: Z-bond, Science China, accepted.

Zhang S, Sun J, Zhang X, Xin J, Miao Q, Wang J, Chemical Society reviews., 2014, DOI: 10.1039/C3CS60409H.

Chen S, Zhang S, Liu X, Wang J, Wang J, Dong K, Sun J, Xu B, Phys. Chem. Chem. Phys., 2014, 16: 5893-5906.

Liu X, Zhao Y, Zhang, Zhou G, Zhang S. J. Phys. Chem. B 2012, 116, 4934–4942

Ionic Liquids – Phase Behavior to Applications

Dr. Mark B. Shiflett, P.E.

DuPont Central Research and Development
Experimental Station
Wilmington, Delaware 19880 U.S.A.

Abstract

The history of ionic liquids (ILs) effectively started in 1914, when the physical properties of ethylammonium nitrate ($[CH_3CH_2NH_3^+][NO_3^-]$, m.p. 13-14 °C) were first reported. ILs are generally defined as salts composed of discrete cations and anions with melting points below 100 °C, and many are liquid at ambient temperature. IL research has grown rapidly over the past decade due to the realization that these materials have many unique properties such as negligible vapor pressure and outstanding solvation potential. Ionic liquids have been further emphasized by the fact that their physical and chemical properties can be finely tuned by varying both the cation and anion.

Our research has focused on accurately measuring vapor-liquid equilibria (VLE) and vapor-liquid-liquid equilibria (VLLE) and using thermodynamic models to understand the phase behavior of binary gas mixtures in ILs. This presentation will focus on the importance of characterizing the global phase behavior of gases in ionic liquids and how this can provide insight into new applications. Solubility measurements of several gases in ILs will be discussed and important experimental details regarding VLE measurements using a gravimetric microbalance and VLLE measurements using a mass-volume technique will be highlighted. VLE data have been successfully correlated with a modified Redlich-Kwong equation of state (EOS), and in certain cases (*e.g.* hydrofluorocarbons) the EOS predicts partial immiscibilities (LLE) with lower critical solution temperatures (LCSTs) in the fluorocarbon-rich side solutions. We have also found that gases such as CO_2 can exhibit different solubility behaviors in ILs (*i.e.* physical and chemical absorption) and that these behaviors can be analyzed with the EOS using a simple association model and excess thermodynamic functions.

Knowledge of gas and IL phase behavior has led to several practical applications including separation of azeotropic gas mixtures, removal of CO_2 from flue gas and new fluid pairs for absorption cooling. These along with other interesting examples from the literature will be discussed.

MECHANISM AND KINETICS FOR CATALYTIC CONVERSION OF CELLULOSE TO ETHYLENE GLYCOL

Junying Zhang, Baolin Hou, Aiqin Wang and Tao Zhang, Dalian Institute of Chemical Physics, Chinese Academy of Sciences, Dalian, China. E-mail: taozhang@dicp.ac.cn

Introduction

Biomass is the only renewable organic carbon in nature. The efficient conversion of biomass into liquid fuels and chemicals is of high importance to the partial replacement of fossil resource with renewable biomass and the decrease of CO_2 emission. Among others, the one-pot catalytic conversion of cellulose into ethylene glycol (EG) is one of the most interesting processes attributed to the high efficiency of tungsten-based catalysts and the huge market of EG (~ 20 million tonnes per year) as the monomer in the plastic industry. This new process was discovered by our group in 2008.[1] Since then, various catalyst formulations have been developed towards the enhancement of the EG yield (EG yield was up to 75%) as well as the understanding of the reaction mechanism.[2-14] Nevertheless, the reaction kinetics is still a great challenge owing to the complicated nature of the consecutive reactions and the difficulties in identifying various intermediates.

The one-pot conversion of cellulose to EG is constituted of three consecutive reactions: (i) the hydrolysis of cellulose to water-soluble celloligosaccharides and glucose, (ii) the C-C cleavage of celloligosaccharides and glucose to produce glycolaldehyde (GA) via retro-aldol reaction pathway, and (iii) the hydrogenation of GA to produce EG over supported transition metal catalysts (typically Ru or Ni). Among them, the second reaction is the key to the whole cascade reactions and it dictates the selectivity of EG to a great extent. Figure 1 shows the reaction mechanism that we proposed based on the temperature-controlled phase-transfer behavior of tungstic acid.

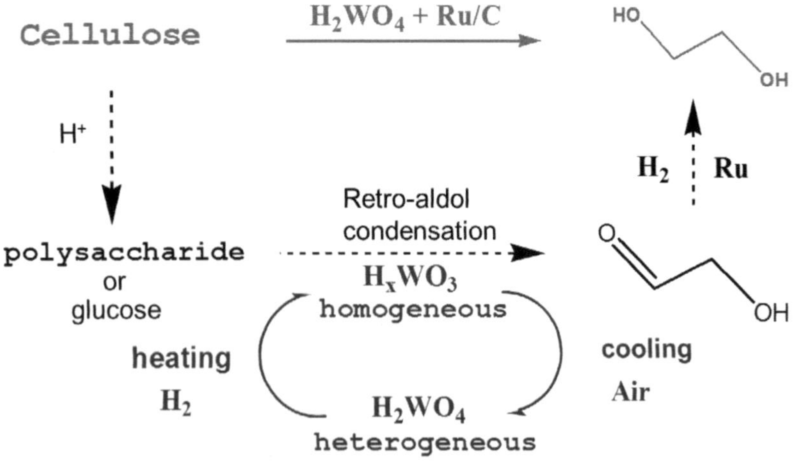

Figure 1. Reaction mechanism proposed for the one-pot conversion of cellulose to EG

In order to get an insightful understanding of the reaction mechanism as well as to optimize the reaction conditions and reactor design, the reaction kinetics of glucose conversion to EG with binary catalyst ammonium metatungstate (AMT) + Ru/C was studied.

Experimental

Reaction

In all experiments, glucose (J & K Chemical), ammonium metatungstate (AMT, Sinopharm Chemical Reagent Co.), and glycolaldehyde (Chemfun Medical Technology (Shanghai) Co., Ltd.) were used as received. Based on our previous study,[14] the glucose or glycolaldehyde transformation reactions were performed in a semi-continuous stainless-steel autoclave (Parr Instrument Company, 300 mL) which is equipped with sampling tube, stirring impeller, and temperature and pressure control system. Typically, 0.05 g AMT was dissolved in 140 mL water and the resulting solution was put into the autoclave. The autoclave was flushed with H_2 five times and then pure H_2 gas was charged until the pressure of 6 MPa. The autoclave was then heated to the desired temperature, into which 10 mL of an aqueous solution of glucose or glycolaldehyde was fed by a Shimadzu LC pump (LC-20A) at a flow rate of 10 mL/min. It took 1 min to finish the feeding process, and this point was considered as the initial time (t = 0). Then, the reaction was started by strong agitation at 1100 rpm. Samples were taken from the reactor at a certain time interval for analysis.

Analysis

After filtration through a 0.45 μm PTFE filter, the liquid samples were analyzed with a high performance liquid chromatograph (HPLC, Aglient 1200) in combination with HPLC-MS, with water as the mobile phase and RI as the detector. For the separation of polyols, a Shodex SC100 column was used with water flow rate of 0.6 mL/min and column temperature of 348 K. For the separation of unsaturated intermediates, a CARBOSep CHO-620 column was used with water flow rate of 0.4 mL/min and column temperature of 353 K. The quantification of intermediates and products was made by an external standard method.

Results and Discussion

Reaction kinetics of the retro-aldol condensation of glucose

The retro-aldol condensation of glucose to glycolaldehyde with AMT as the catalyst is the key reaction for the conversion of cellulose to EG. Through reasonable assumptions, the reaction is simplified into three consecutive reactions: retro-aldol condensation of glucose to erythrose and glycolaldehyde (R1), retro-aldol condensation of erythrose to two moles of glycolaldehyde (R2), and conversion of glycolaldehyde to side products (R3). Fitting of the experimental data shows that the apparent activation energies for R1, R2, and R3 are 141.3, 79.9, and 52.7 kJ/mol, respectively, while the reaction orders are 1st, 1.7^{th} and 2.5^{th}, respectively. On the other hand, the reaction rate of R1 is 0.257^{th}-order dependence on the concentration of AMT catalyst. The high activation energy of R1 suggests that a high temperature is favorable to the formation of glycolaldehyde, and the low activation energy and the high reaction order of R3 suggests that glycolaldehyde is very easy to transform into side products under the reaction conditions. Therefore, in order to achieve a high EG yield, high reaction temperatures and low concentration of glycolaldehyde are required.

The rate constants at different temperatures and Arrhenius parameters of the above three reactions are used for validation of the kinetic models. Figure 2 shows the experimental data obtained at different initial concentrations of glucose (11.2 mol/m^3, 16.5 mol/m^3) as well as the calculation results based on the kinetic parameters. For the glucose concentration with time course, good agreement was found between predicted and experimental results. However, for the glycolaldehyde concentration with the reaction time, it can be seen that the predicted maximum glycolaldehyde concentration is higher than the experimental values. One possible reason for this disagreement is the simplification in the kinetic model establishment. It is known that the isomerization of glucose to fructose always occurs under the reaction conditions although to a less significant extent, and the subsequent retro-aldol reaction of fructose takes place too. In our model simplification, the two reactions are not considered, which would lead to over-estimation of the glycolaldehyde production. Nevertheless, our simplified model could well predict the overall trend of the target product glycolaldehyde, which is indeed encouraging given that the overall reaction from glucose to glycolaldehyde is rather complicated. The experimental and predicted results at different temperatures (463 K and 473 K) also indicate that better agreement is obtained for the glucose concentration than glycolaldehyde with the time course, but the overall trend can be well predicted, validating the three kinetic models.

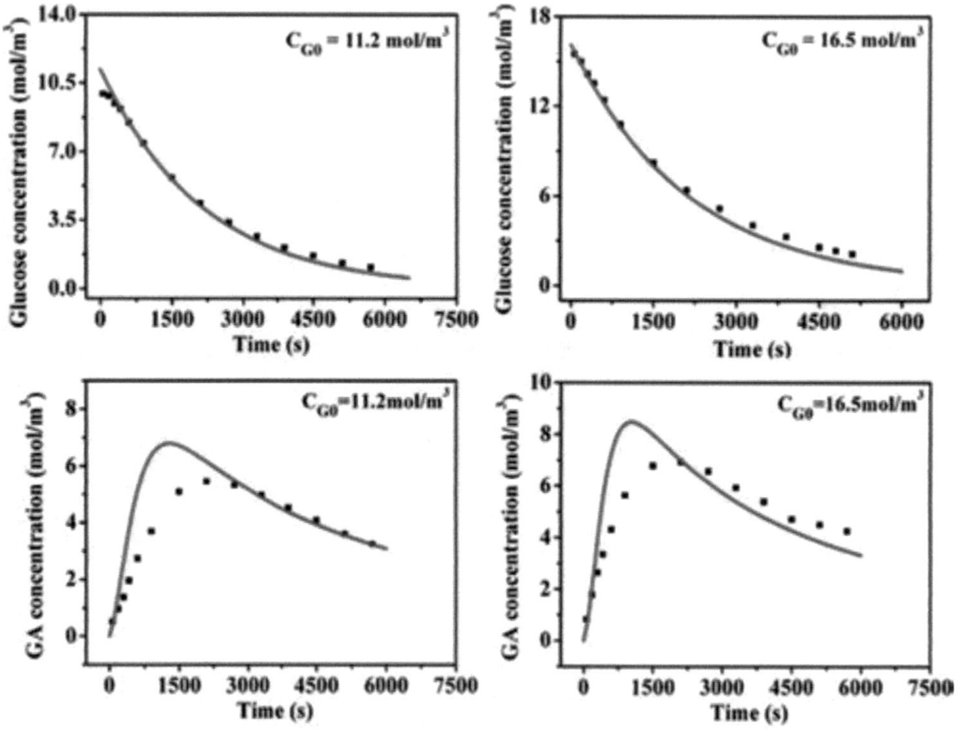

Figure 2. Experimental (dots) and calculated (lines) results for the overall reaction of glucose at different initial concentration of glucose. T = 433 K.

Reaction kinetics of glucose and glycolaldehyde hydrogenation

By assuming dissociative adsorption of H_2 on Ru/C surface and non-competitive adsorption of hydrogen with glucose or glycolaldehyde, we use LHHW model to describe the kinetics of glucose and glycolaldhyde hydrogenation with or without the presence of AMT. The result shows that the apparent activation energies are 49.6 and 42.6 kJ/mol for glucose hydrogenation and glycolaldehyde hydrogenation, respectively, and the adsorption enthalpies of glucose and glycolaldehyde on Ru/C catalyst are comparable (16.5 vs. 10.3 kJ/mol). However, the rate constant of glycolaldehyde hydrogenation is 4 times faster than that of glucose, which ensures the preferential hydrogenation of glucose when the two substrates are co-present. It has been for the first time found that the presence of AMT retarded significantly the hydrogenation of both aldoses, and the reaction orders were -0.69 and -0.24 with respect to AMT concentration for glucose and glycolaldehyde hydrogenation, respectively, manifesting the suppressing effect was more pronounced on the glucose hydrogenation. By considering the complexing interaction between AMT and aldoses, the kinetic model without AMT was modified and the modified model could well predict the experimental data in the presence of AMT, as shown in Figure 3.

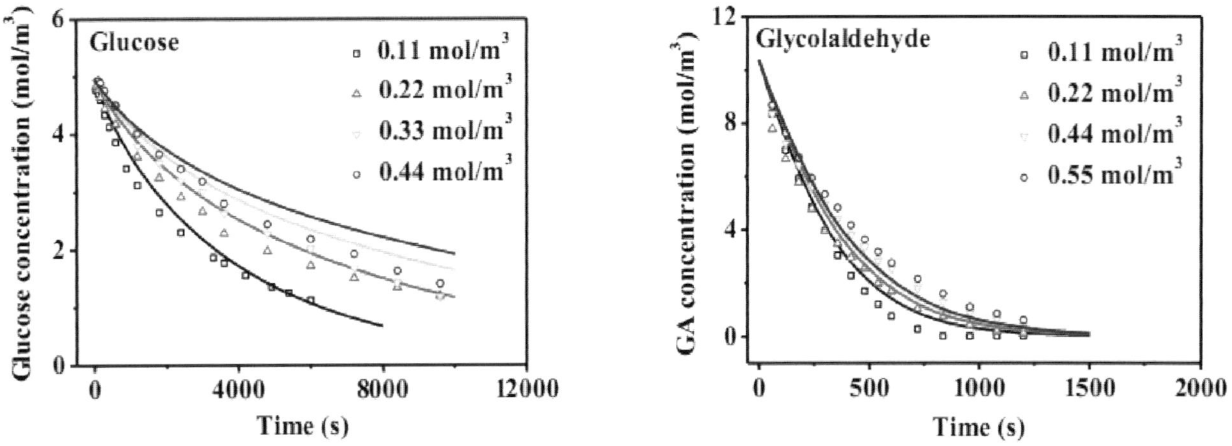

Figure 3. Experimental (dots) and predicted (lines) concentration profiles of glucose and glycolaldehyde at different AMT concentrations. Reaction conditions: $C_{G0} = 4.96$ mol/m^3, $C_{GA0} = 10.4$ mol/m^3, T = 403 K, $P_H = 6$ MPa, 0.2 g 1%Ru/C.

The kinetics of glucose conversion to EG

Based on the kinetic parameters of glucose retro-aldol condensation and glucose and glycolaldehyde hydrogenation, the whole reaction kinetics of glucose conversion to EG has been reached, and the EG and sorbitol yields with temperature/time during the reaction could be well predicted, as shown in Figure 4.

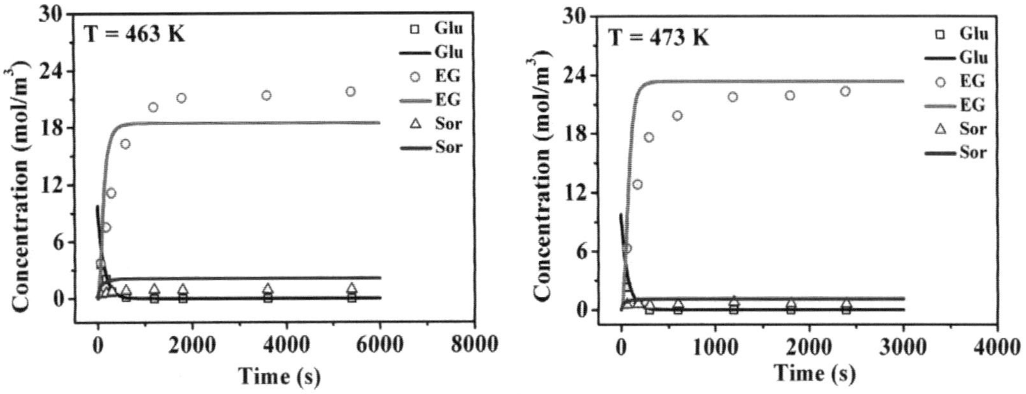

Figure 4. Modeling results for ethylene glycol production from glucose over AMT and Ru/C at different temperatures.

The effect of tungstic compounds on the hydrogenation of glucose

On the basis of the above kinetic studies, we further investigated various acids and tungstic compounds on the hydrogenation of glucose. The results showed that HCl, H_2SO_4, H_3BO_3, H_3PO_4, and HNO_3 had negligible effect, while all the tungstic compounds imposed negative effects on the hydrogenation of glucose over Ru/C catalyst, and the suppressing effect followed the order of H_2WO_4> HPW> WO_3> AMT> HSiW, as shown in Figure 5. This order is basically the same as the order of EG yields in the reaction of glucose conversion, suggesting the important role of competition between glucose hydrogenation and retro-aldol condensation in controlling the selectivity of EG.

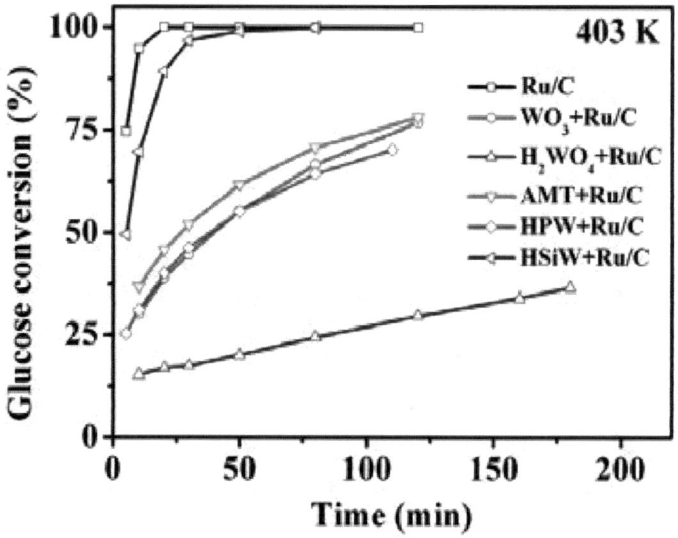

Figure 5. Glucose conversion versus reaction time with different catalysts. Reaction conditions: 0.375 mmol of W, 0.30 g 1%Ru/C, 6 MPa H_2, stirring speed 1000 r/min, 403 K.

Conclusion

The one-pot conversion of cellulose to EG consists of three consecutive reactions, which necessitates the employment of a dual catalyst composed of a tungstic compound and a hydrogenation metal. Our work revealed that tungstic compounds effectively and uniquely catalyze the retro-aldol condensation of glucose to selectively produce glycolaldehyde, while Raney Ni and Ru/C are active and durable catalyst for the subsequent hydrogenation of glycolaldehyde to EG. Meanwhile, both Raney Ni and Ru/C catalysts are also able to catalyze the reaction of glucose hydrogenation to sorbitol that will compete with the reaction of glycolaldehyde hydrogenation resulting in the decrease of EG selectivity.

In the present work, the kinetics of one-pot conversion of glucose EG was studied using a batch reactor. The retro-aldol condensation of glucose to glycolaldehyde under the catalysis of AMT is a key reaction for the overall transformations, and the high apparent activation energy (141.3 kJ/mol) of this reaction indicates that high temperature is favorable to the formation of glycolaldehyde, while the concentration of glycolaldehyde in the reactor should always be kept at a very low value in order to avoid the condensation side reaction. The reaction rate of glycolaldehyde hydrogenation was much faster than that of glucose hydrogenation, which led to the preferential hydrogenation of glycolaldehyde when glucose was co-present with glycolaldehyde in the reaction system. However, in the presence of AMT, it was found that AMT significantly reduced the hydrogenation reaction rate of aldoses (glucose or glycolaldehyde), and this suppressing effect was more pronounced for glucose hydrogenation. The kinetic results have provided insightful understanding of the dual effects of the tungstic compound in the one-pot conversion of glucose or cellulose to EG, i.e., on one hand it catalyzes the selective cleavage of C-C bonds of glucose to produce glycolaldehyde, and on the other hand it inhibits the hydrogenation of aldoses especially that of glucose by poisoning Ru/C surface. Finally, by combining the two kinetic models, i.e., retro-aldol condensation of glucose and hydrogenation of glucose/glycolaldehyde with the dual catalyst AMT + Ru/C, the overall kinetic model for the one-pot conversion of glucose to EG was developed. Our preliminary results have shown that both EG and sorbitol yields from glucose conversion can be predicted well with this overall kinetic model.

References

1. N. Ji, T. Zhang, M. Zheng, A. Wang, H. Wang, X. Wang, J. Chen, *Angew. Chem. Int. Ed.* **47**, 8510-8513 (2008).
2. Y. Zhang, A. Wang, T. Zhang, *Chem. Commun.* **46,** 862-864 (2010).
3. M. Zheng, A.Wang, N. Ji, J. Pang, X. Wang, T. Zhang, *ChemSusChem* **3**, 63-66 (2010).
4. J. Pang, A. Wang, M. Zheng, T. Zhang, *Chem. Commun.* **46,** 6935-6937(2010).
5. J. Pang, A. Wang, M. Zheng, T. Zhang, *Ind. Eng. Chem. Res.* **50,** 6601-6608 (2011).
6. N. Ji, M. Zheng, A. Wang, T. Zhang, J. G. Chen, *ChemSusChem* **5,** 939-944 (2012).
7. Z. Tai, J. Zhang, A. Wang, M. Zheng, T. Zhang, *Chem. Commun.* **48,** 7052-7054 (2012).
8. L. Zhou, A. Wang, C. Li, M. Zheng, T. Zhang, *ChemSusChem* **5,** 932-938 (2012).
9. J. Pang, A. Wang, M. Zheng, Y. Zhang, Y. Huang, X. Chen, T. Zhang, *Green Chem.* **14,** 614-617 (2012).
10. C. Li, M. Zheng, A. Wang, T. Zhang, Energy Environ. Sci. 5, 6383-6390 (2012).
11. G. Zhao, M. Zheng, J. Zhang, A. Wang, T. Zhang, Ind. Eng. Chem. Res. 52, 9566-9572 (2013).
12. Z. Tai, J. Zhang, A. Wang, J. Pang, M. Zheng, T. Zhang, *ChemSusChem* **6,** 652-658 (2013).
13. A. Wang, T. Zhang, *Acc. Chem. Res.* **46,** 1377-1386 (2013).
14. X. Zhao, J. Wang, C. Chen, Y. Huang, A. Wang, T. Zhang, *Chem. Commun.* **50,** 3439-3442 (2014).

390028 Glycerol Carbonate from Glycerol and Urea Using Heterogeneous Catalysis

Abstract:

Colombia is a producer biodiesel country mainly by transesterification and several investigations had been carried out in the search for new uses for the glycerin as raw material. Glycerol carbonate is a neoteric solvent with diverse uses as intermediate species for other chemical production. In the present work the production of glycerol carbonate using glycerin and urea was investigated, using heterogeneous catalysis. The several reactions used to obtain glycerol carbonate was evaluated to determinate the atoms economy, a preliminary analysis of the chemical fate of the reactants and products was carry out using EPIWEB 4.1.
Two types of catalysts were tested: supported catalysts (CuO/ Al2O3), distribution and ZnO/ Al2O3, both with shell egg distribution. The catalytic support and the final catalysts were characterized by area BET, TGA, FTIR The catalytic test were made in a batch reactor varying the urea/ glycerin ratio and temperature. A experimental design was used in this research.

387399 Cost Effective Lipid Production By Lipomyces Starkeyi Using Different Carbon Sources

Abstract:

Energy demand and environmental concerns are making it essential to find renewable and sustainable alternatives to fossil fuels. Biofuel can supplement diesel fuel in an environmentally friendly and economical way as long as the raw material employed is of low cost and can be derived from sustainable sources. Such an alternative feedstock, from non-vegetable source, could be provided by microbial lipid produced by oleaginous microorganisms which has chemical composition similar to that of vegetable oils. The oleaginous yeast, *Lipomyces starkeyi*, can be cultivated on different carbohydrates for production of over 75% of cell dry weight as lipids. Over one billion tons of lignocellulosics can be available in the USA annually which could be used for microbial lipid production. Several byproducts are formed during the hydrolysis of lignocellulosics. These products include pentoses and hexoses (furfural, hydroxymethyl furfural), phenolic compounds originating from lignin (hydroxybenzaldehyde, syringaldehyde) and organic acids. The concentration ranges of these byproducts in the hydrolyzates depend on the type of lignocellulosics as well as the method of hydrolysis employed. Out of the byproducts produced, furfural, HMF, and acetic acid are the key components which inhibit the microbial growth. This research work deals with the synthetic hydrolyzate compositions in shake flasks to identify tolerance limit of the byproduct (furfural, hydroxymethyl furfural, acetic acid) concentrations in the medium by *Lipomyces starkeyi* cells. Response surface methodology will be utilized to design and analyze the experiments. All the experiments will be conducted in 500 mL shake flasks with the working volume of 125 mL. The cell mass and lipid content are analyzed by dry weight and nile-red fluorescence respectively. The glucose concentration is measured by HPLC analysis.

392767 Membrane Reactor for Scalable Manufacturing of Nano-Material Products

Abstract:

Nano-sized solid particles with tailored properties are desired for a variety of applications from pharmaceutical, material, chemical, environmental to energy industry. However, synthesis of the nano-particles with design properties in a scalable and sustainable way has been a major challenge. In this presentation, the author will introduce membrane reactor technologies for continuous production of nano-particles through multiphase reactions. The literature progress will be brief reviewed, including preparation of nano-particles for controlled drug delivery. The recent progress made by this research group for production of transition metal (Fe, Co, Ni) nano-particles will be reported.

392870 Green Technologies to Decrease Water and Energy Consumption in the Laundry Process

Abstract:

Home laundry process results in among the largest water and energy consumption annually on our planet. The past decade has experienced a rapid pace of green innovations in both durable and consumable cleaning products, many of which are delivering improved consumer benefits as well. This presentation will start off by discussing the water and energy decrease trends within laundry appliances over the past two decades and their impact on the environment. A deeper dive into a few of the more impactful technologies that has enabled this progress will be presented including sensing technologies timed dispensing of cleaning additives, adapting mechanical forces to improve performance of specific cleaning components, use of non-traditional solvents as cleaning medium beyond water. Finally the presentation is going to wrap with a discussion of energy efficiency laundry trends across the globe and highlight opportunities between the appliance manufacturers and chemical industry to together drive and shape these trends for global energy efficient cleaning technologies.

379398 Filtration and Decontamination of Toxic Chemicals and Ambient Contaminants By the Multifunctional Zirconium Hydroxide

Abstract:

Novel sorbent development remains an active area of research in a variety of industries. In respiratory and collective protection, efforts mainly focus on new materials for enhanced and broadened protection against toxic chemicals. Traditionally, activated, impregnated carbons have been used to provide efficient protection against chemical warfare agents and many toxic industrial chemicals. However, these materials are typically over-designed for persistent chemical agents, while high volatility, reactive gases are removed less efficiently.

In this talk, we discuss the paradigm-shifting development of a reactive substrate targeting acidic/acid-forming gases. We explore the development of zirconium hydroxide, a substrate with surface hydroxyl groups that provide efficient removal of chemicals such as hydrogen chloride and sulfur dioxide, among others. Additives such as cobalt enhance reactivity towards nitrogen dioxide. Together, not only are highly efficient respiratory and collective protection filters possible, potentially reducing the filter volume, but new components can be developed. One such component is a composite fabric/zirconium hydroxide pre-filter that can remove ambient contaminants such as NO_x and SO_x, reducing quantities approaching the filter and therefore prolonging filter life.

In addition to toxic industrial chemicals and ambient contaminants, zirconium hydroxide also possesses extraordinary reactivity towards persistent chemical warfare agents such as V and G agents. New techniques are under development for on-demand decontamination of these agents. The multi-faceted nature of zirconium hydroxide results in a material with ever-changing capabilities allowing for enhanced protection.

392822 Transition Metal-Ceria Nano-Composite Catalyst for Air Pollution Control Application

Abstract:

In this presentation, I will discuss the design of nano-composite catalysts by utilizing interfacial interactions of two different kinds of materials to generate strong synergistic effects. The design principle will be illustrated with transition metal-ceria metal oxide composites for oxidative destruction of environmental pollutants. The strong synergism is demonstrated for oxidation of CO, hydrocarbons, and soot. In particular, the latest progress for low-temperature reduction of NO will be reported.

366433 Automotive & Diesel Emissions Control Technologies

Abstract:

Automotive & Diesel Emissions Control Technologies

Steven B. Ogunwumi

Corning Incorporated, Crystalline Materials Research, Corning NY 14831, USA

Abstract

Nitrogen oxides (NOx), Carbon monoxide, (CO) Hydrocarbons (HC) and Particulate matter (PM) from gasoline and diesel emissions pose serious health hazards and environmental issues. These emissions have been implicated in acid rain, ozone and urban smog formation. NOx and CO gases are considered to be highly toxic, and PM and some HC may be carcinogenic.

As a result, legislation continues to enforce stringent regulation of these emissions. The development of the three way catalyst (TWC) has helped to mitigate this problem with gasoline vehicles. The standard gasoline combustion engine operates in a stoichiometric fuel/air regime. Thus, a conversion efficiency of more than 95% NOx to nitrogen can be attained by operating near or close to stoichiometry. However, due to the lean-burn (oxygen-rich) conditions of diesel exhausts, a conventional TWC technology solution cannot be applied in a diesel exhaust.

This presentation provides an overview of the after-treatment technology available for automotive and diesel exhaust emission remediation. Information is provided on the development of the catalyst substrate, diesel filter, catalyst technology, and emissions chemistry.

357606 Removal of Low Concentration Radioactive Noble Gases Using Metal-Organic Frameworks-a Column Breakthrough Study

Abstract:

Metal-organic frameworks (MOFs) are novel porous materials which have potential applications in gas adsorption and separation. Radioactive isotopes of xenon (Xe) and krypton (Kr) exist in the off-gases from nuclear power plants and they need to be removed to prevent radiation pollution, especially Kr due to its long half-life isotope ^{85}Kr. Two metal-organic framework materials were investigated to determine the removal efficiency and capacity of MOF materials for krypton removal from air at non-cryogenic temperatures. Our two bed breakthrough measurements on the Ni/DOBDC and a partially fluorinated FMOF-Cu indicate these MOF materials can capture and separate parts per million levels of Xe and Kr from air. Moreover, the removal efficiencies and adsorption capacities for Kr on these two MOFs were further increased after Xe was removed in the first bed to reduce competitive adsorption. In addition, the FMOF-Cu showed a Kr adsorption capacity of 1.03 mmol/kg at 40 ppm and 233 K, which is one of the highest values reported in the literature. This extraordinary capacity is possibly due to a temperature gate effect known to the FMOF-Cu. Our results show a promising future for MOFs in a radioactive nuclides separation from spent fuel.

392871 Novel Porous Carbon Adsorbents for CO2 Capture from Flue Gas and Natural Gas Upgrade

Abstract:

A few N-functionalized porous carbon adsorbents were synthesized and characterized for capturing carbon dioxide from flue gas and separating carbon dioxide from natural gas. The hydroquinone and quinone functional groups were grafted into a hierarchical porous carbon framework via Friedel-Crafts reaction for selective capture and removal of carbon dioxide from flue gases and natural gas. Adsorption isotherms of CO_2, CH_4, and N_2 were measured at 1 atm and three different temperatures, and used for predicting the carbon adsorbents' separation capability by using the Ideal Adsorbed Solution Theory (IAST). At 298 K and 1 atm, the highest CO_2/N_2 selectivity value of 26.5 is obtained on the hydroquinone-grafted carbon (OAC-1), which is significant improvement (158.7%) of the pristine porous carbon (OAC-0), and the highest CO_2/CH_4 selectivity value of 4.6 is obtained on the quinone-grafted carbon (OAC-2), 28.4% better than the pristine porous carbon. The CO_2 adsorption capacity on the modified activated carbon was increased from 3.02 mmol g^{-1} to 3.46 mmol g^{-1} at 298 K and 1 atm, although the specific surface area was reduced after modification. In addition, transient breakthrough simulations for $CO_2/CH_4/N_2$ binary mixtures were conducted to demonstrate the good separation performance in a fixed bed adsorber. The heats of adsorption of carbon adsorbents are lower than other adsorbents, suggesting a low regeneration energy requirement. Overall the new porous carbon adsorbents developed in this work look very promising for flue gas treatment and natural gas upgrading.

Nanoporous Carbon from Renewables for Supercapacitor Applications

Satish K. Nune, Ju-Won Jeon, Libing Zhang, John P. Lemmon, Melvin P. Tucker, Peter B. McGrail, Bin Yang

Due to the increased energy consumption, efficient and cost-effective electrical energy storage systems are highly desired. Supercapacitors that bridge the gap between batteries (high energy and low power) and capacitors (low energy and high power) have been playing crucial role as one of the energy storage systems and have started to realize their potential in transportation as well as consumer electronics. However, their use is relatively limited because of poor performance (low energy density) and high cost. Significant efforts have been made to develop highly efficient less expensive supercapacitors. We proposed to use low-cost renewable biomass such as lignin as a potential precursor to produce porous carbons for Supercapacitor applications. Lignin is a natural amorphous polymer that acts as an essential glue to give plants their structural integrity. It is a main component in lignocellulosic biomass (15-30% by weight, 40% by energy), together with cellulose and hemicelluloses. Simple carbonization of lignin results in carbons with very low surface area that is not suitable for supercapacitors. We investigated the effect of molecular weights of the lignin precursors on surface area and energy densities. We will present our results on the carbon residues obtained from three different lignin precursors that differ in molecular weights and their electrochemical performances.

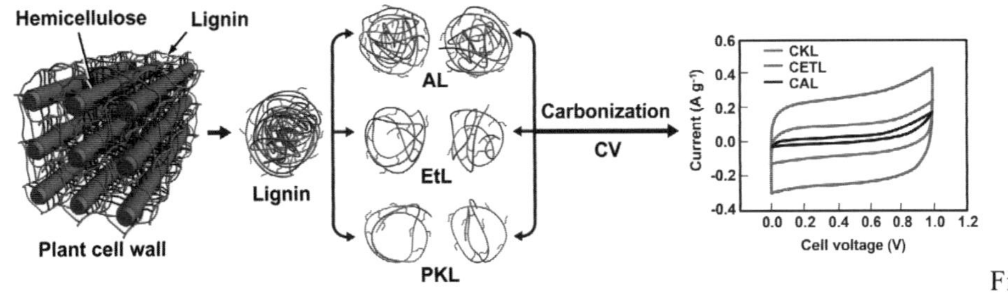

Figure 1: Extraction of ligin with three different molecular weights and corresponding CV plots.

Computational scheme for design of ionic liquid in extraction of pharmaceutical intermediate

Madakashira Harini[†], Shanu Jain[†], Jhumpa Adhikari[†], Santosh B. Noronha[†] and K. Yamuna Rani[*]

[†]*Department of Chemical Engineering, Indian Institute of Technology Bombay, Mumbai-400076*
[*]*Chemical Engineering Division, Indian Institute of Chemical Technology, Hyderabad-500007*

Ionic liquids (ILs) are green solvents that have tuneable properties and are considered as potential alternatives to volatile organic compounds in many industrial applications. Whenever, there is a need for replacement of a non-green solvent in a process due to environmental, health and toxicological concerns; ILs can be an alternative. The properties of ILs can be optimized for a specific task by the combination various cations and anions; and also by varying alkyl chain length or by addition of a functional group on the cation/anion. As it is tedious and uneconomical to conduct experiments on various ILs to find its suitability for specific process, computer aided molecular design[1] (CAMD) methodology has been adopted to design an IL for efficient extraction of R-phenylacetylcarbinol (PAC), from aqueous phase. PAC is an intermediate in the manufacture of several pharmaceutical products and is currently produced by means of whole cell biocatalysis.

The properties that are of interest, for extraction of PAC, include solvent power, selectivity, distribution coefficient, solvent loss, octanol-water partition coefficient, melting point (T_m), viscosity and thermal decomposition temperature (T_d). These properties, except the T_m, visosity and T_d, are ratios of infinite dilution activity coefficient values and are estimated using the universal functional activity coefficients (UNIFAC) method;[2] while the T_m and T_d are estimated using a group contribution model. A basis set of 16 ILs has been chosen, to which an alkyl chain or a functional group (with a maximum of four groups) can be attached at two or three different positions, based on the structure of the IL. A total of 23 groups (including repetitions) have been chosen, whose combinations lead to an attachment of hydroxyl, alkyl, vinyl and ketone based ILs.

A variant of the genetic algorithm, namely, non-dominated sorting genetic algorithm (NSGA-II)[3], and the direct search method, have been employed to arrive at list of feasible ILs that can extract PAC efficiently. The structural constraints have been incorporated to generate a viable ionic liquid. The bounds on all the target properties are specified based on the property values of toluene; as current industrial practice employs toluene as the solvent. A multi objective function that minimizes solvent loss and maximizes selectivity, distribution coefficient and solvent power, has been utilized in optimization to generate the list of feasible ILs. Among the ILs considered in this study, the ILs containing anions Tf_2N^- and FAP^- have been determined as the best solvents for extraction of PAC. The resultant molecules obtained from the direct search method and NSGA-II are in excellent agreement and yielded 2-methyl 2-butenyl [N-C3OHPY][FAP]

(1-(3-hydroxypropyl) pyridinium tri(pentafluoroethyl) trifluorophosphate) as the best solvent. The order of extraction capability for the side chain attachment in this case study suggests, vinyl > alkyl > ketone > alcohol; and among alkyls, butyl > propyl > ethyl > methyl. This observation on alkyl groups is in agreement with the experimental trend of increase in hydrophobicity with chain length. In addition, some of the experimental findings on specific ILs (with Tf_2N^- and FAP^- anions) reported in literature[4] are in agreement with our CAMD findings.

References:

1. Karunanithi, A. T. & Mehrkesh, A. *AIChE J.* **59,** 4627-4640 (2013).
2. Lei, Z., Dai, C., Liu, X., Xiao, L. & Chen, B. *Ind. Eng. Chem. Res* .**51**, 12135-12144 (2012).
3. Deb, K., Pratap, A., Agarwal, S. & Meyarivan, T. *IEEE Trans.Evolutionary Computation* **6,** 182–197 (2002).
4. Marciniak, A. & Karczemna, E. *Fluid Phase Equilib.* **304,** 121–124 (2011).

Hydrogen Bond Promoted Conversion of Carbon Dioxide to Cyclic Carbonates over Ionic Liquids Catalysts

Suojiang Zhang*, Weiguo Cheng

Key Laboratory of Green Process and Engineering, Institute of Process Engineering, Chinese Academy of Sciences, P.O. Box 353, Beijing, 100190, China

**Corresponding author: sjzhang@ipe.ac.cn*

Carbon dioxide is believed to be responsible for global warming and climate change, while it is also a nontoxic, abundant, inexpensive and sustainable C1 resource. Chemical fixation of CO_2 into cyclic carbonates has attracted extensive attention in both academia and industry due to 100% atom-economical reactions (Scheme 1). Although the cycloaddition of CO_2 with epoxides into cyclic carbonates catalyzed by ionic liquids(ILs) has been studied extensively, the tunable structures of the ILs by combination of different cations with anions is still desirable. Currently, it is found that hydrogen bond donors (HBD) could activate the epoxides to enhance the activity of ILs. However, the detailed effect of hydrogen bond on the reaction is unclear.

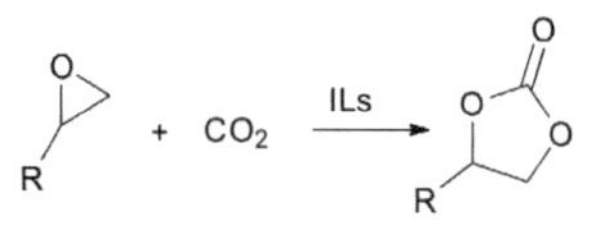

Scheme 1 Fixation CO_2 and epoxides into cyclic carbonates

In this work, we investigated effect of hydrogen bond on the cycloaddition reaction through experimental and theoretical studies[1-5]. The results show that HBD could remarkably reduce the activation energy and promoting ring-opening of epoxide. The cations and anions of ionic liquids show a synergistic effect on the cycloaddition reactions. The effect of hydrogen bond on the reaction is different according to the nature of functional groups in cations. Based on understanding of the reaction mechanism, a series of ILs catalysts with hydrogen bond groups were designed for fixation of CO_2. This work reported here provides a molecular level understanding of effect of hydrogen bond on the reaction and provide the basis of rational design of ILs.

Keywords: hydrogen bond, ionic liquid, carbon dioxide, cyclic carbonate, catalyze

References

1. Cheng WG, Chen X, Sun J, Wang JQ, Zhang SJ. SBA-15 supported triazolium-based ionic liquids as highly efficient and recyclable catalysts for fixation of CO_2 with epoxides. *Catalysis Today*, 2013,200: 117- 124
2. Wang JQ, Sun J, Cheng WG, Dong K, Zhang XP, Zhang SJ. Experimental and theoretical studies on hydrogen bond-promoted fixation of carbon dioxide and epoxides in cyclic carbonates. *Phys. Chem. Chem. Phys.* 2012, 14: 11021- 11026.
3. Sun J, Han LJ, Cheng WG, Wang JQ, XP, Zhang SJ. Efficient Acid–Base Bifunctional Catalysts for the Fixation of CO_2 with Epoxides under Metal- and Solvent-Free Conditions. *ChemSusChem.,* 2011, 4(4): 502-507.
4. Dong K, Song YT, Liu XM, Cheng WG, Yao XQ, Zhang SJ. Understanding Structures and Hydrogen Bonds of Ionic Liquids at Electronic Level. *J. Phys. Chem.B. J.* 2012, 116: 1007-1017.
5. Sun J, Yao XQ, Cheng WG, Zhang SJ. 1,3-Dimethylimidazolium-2-carboxylate: A zwitterionic salt for the efficient synthesis of vicinal diols from cyclic carbonates. *Green Chem.*, DOI: 10.1039/C4GC00279B.

380252 The Development of Anti-Inflammatory Drugs As Ionic Liquids for Topical Drug Delivery

Abstract:

Ionic liquids (ILs) are organic salts whose melting points are below 100°C and are composed entirely by ions. Due their physicochemical properties, they have been used in wide variety of applications such as green solvents in chemical syntheses and separations. Most recently, ionic liquids have attracted a great deal of attention due the large number of possible cation-anion combinations of active pharmaceutical ingredients (APIs) that can produce ionic liquids with biological activities. Several ionic liquids can be in a liquid state at room temperature and some of their benefits include improved solubility, stability, bioavailability and alternative ways for drug delivery. These room temperature ionic liquids (RTILs) are very attractive candidates for topical drug delivery, since there is no need to add any solvent to the active pharmaceutical drug. In addition, the lipophilicity of the ionic liquid can be tuned by proper selection of the counterion and therefore it will facilitate the absorption of the drug into the skin. Here, we will report the use of several synthesized room temperature ionic liquids (RTILs) from active pharmaceutical ingredients for topical drug delivery. The RTILs synthesized incorporate an anti-inflammatory drug with different GRAS (Generally Regarded as Safe) and inactive ingredients used in pharmaceutical formulations. Anti-inflammatory drugs are used in the treatment of mild to moderate pain and inflammation. The most serious side effects of anti-inflammatory drugs are caused by the systemic absorption of the drug and can include kidney failure, liver failure and ulcers. Our goal is to eliminate these serious side effects by employing a topical anti-inflammatory drug in the form of an ionic liquid. The RTILs were synthesized through a salt metathesis reaction and they were characterized by spectroscopic and thermoanalytical techniques (i.e. NMR, DSC). Drug permeability experiments were also performed to determine their transport ability through the skin. Details about the synthesis, properties and permeability experiments of the RTILs will be discussed.

385168 Integrated Low Temperature Ionic Liquid Pretreatment and Fractionation of Biomass

Abstract:

The high crystallinity of cellulose of native lignocellulosic biomass, while imparting structural integrity and mechanical strength to the material, renders it recalcitrance towards hydrolysis aimed at producing glucose. Neither water molecules nor catalysts for hydrolysis (saccharification) are able to easily penetrate the crystalline matrix. Room temperature pre-processing is coupled with ionic liquid incubation (IL) at low temperatures ranging between 30 and 50°C, producing an easily hydrolysable cellulosic substrate at dramatically lower incubation temperatures. Moreover, by altering the IL displacement solvents, this method can be tuned to retain or reject biomass components from the IL phase, providing a method for separation of hexose and pentose sugars.

362343 Using Glycerol As a Co-Solvent for Ionic Liquid Pretreatment of Rice Hulls

Abstract:

Over 100 million tons of rice hulls (husks) are generated each year at rice processing centers around the world. These cannot be easily composted and often end up in landfills. They can be pretreated with ionic liquids (ILs) prior to enzymatic hydrolysis to produce glucose or xylose, which can be fermented for biofuel production. However, the present high cost of ILs is a deterrent to large scale use. The non-chloride containing ILs 1-ethyl-3-methylimidazolium acetate, 1-ethyl-3-methylimidazolium formate, 1,3-dimethylimidazolium dimethylphosphate, and 1-ethyl-3-methylimidazolium diethylphosphate diluted with glycerol were investigated for rice hull (husk) pretreatment. When diluted with up to 75% glycerol and used for a 110 °C, three hour pretreatment, 1-ethyl-3-methylimidazolium formate was found to give glucose and xylose yields after enzymatic hydrolysis similar to those of pure 1-ethyl-3-methylimidazolium formate. Diluting with glycerol was also found to decrease solvent viscosity and promote solvent recovery. Such findings suggest that glycerol dilution in IL-biomass pretreatment may, in some cases, enhance pretreatment.

Application of Ionic Liquids in the Cleaner Process of Isobutane Alkylation

Guoying Zhao, Haifang Li, Qian Huang, Zhimao Zhou, Suojiang Zhang*
Institute of Process Engineering, Chinese Academy of Sciences,
100190, Beijing, PR China

*Corresponding author: sjzhang@ipe.ac.cn

Alkylates are viewed as the highest-quality gasoline blending components and usually called as " liquid gold" due to their high octane number in combination of benign (eco) toxicology.[1] There has been increased demand in alkylate produced using low molecular weight hydrocarbons from petroleum refinery, biorefinary and coal refinary process in recent year. The global alkylate production capacity reached up to 83 milliontons. However the established commercial process to produce alkylate are not environmental friendly at all, which typically use sulfuric acid or hydrogen fluoride as catalyst. A cleaner, safer and more efficient alkylation process needs to be developed to keep alkylate competitive with other gasoline blending components. In recent year, aicidic ionic liquids, expecially chloroaluminate ionic liquids, have been investigated to be alternative catalysts for isobutane alkyaltion. We have developed a series of non-chloroaluminate ionic liquids to improve the isobutane alkylation process cleaner.[3,4] Effect of the nature and composition of cation and anion of the ionic liquids on their acidity, catalytic activity and selectivity, and the optimal alkylation process parameters, have been investigated in detail. It was found that the ions designability and synergia give the ionic liquids the outstanding catalytic performance in isobutane alkylation, which are mainly shown in the following several points: 1) Higher-quality alkylate with desired TMP up to 80.1% and RON up to 97.2, was obtained using composite ionic liquids than sulfuric acid at the same conditions. 2) The ionic liquids with the chosen anion and cation can be used as a "buffer" to resist the acidity change of the composite ionic liquids, and/or promote the reaction rate of hydride transfer from isobutane to the intermediate cation, which are limiting step for isobutane alkylation, and so on. 3) The usage of the composite ionic liquids can reduce down to 10 vol% that of sulfuric acid to catalyze effectively isobutane alkylation.

Keywords: ionic liquid, isobutane alkylation , alkylate, cleaner process

Reference

[1] Ren HL, Zhao GY, Zhang SJ, Cui P, Huang J. Triflic acid catalyzed isobutane alkylation with trifluoroethanol as a promoter. Catalysis Communications. 2012; 18: 85–88

[2] Xing XQ, Zhao GY, Cui JZ. Chlorogallate(III) ionic liquids: Synthesis, acidity determination and their catalytic performances for isobutane alkylation. Science China: Chemistry. 2012; 55: 1542–1547

[3] Xing XQ, Zhao GY, Cui JZ, Zhang SJ. Isobutane alkylation using acidic ionic liquid catalysts. Catalysis Communications. 2012; 26: 68–71

[4] Cui P, Zhao GY, Ren HL, Huang J, Zhang SJ. Ionic liquid enhanced alkylation of iso-butane and 1-butene. Catalysis Today. 2013; 200: 30–35

383550 Imparting Ionic Liquid Character to Commodity Chemicals

Abstract:

We have observed that certain bulky salts that are commonly used to synthesize ionic liquids are also highly soluble in a number of polar organic solvents. Thus, ionic liquid character can be easily imparted to commodity chemicals in a straightforward manner. Such mixtures provide a tunable alternative reaction media/solvent, and we have observed that this approach can have significant influence on the properties and yield of the desired product. Furthemore, a mixing approach can streamline/simplify the production of ionic liquids themselves and eliminate separation/purification steps. In this presentation, we will detail thermophysical properties of these polar organic + salt mixtures and discuss the improvements they provide with respect to chemical reactions and separations.

Pyrolysis and Char Gasification Studies of Municipal Solid Waste (MSW)

Sireesha Aluri[1], Pradeep K. Agrawal[1], John D. Muzzy[1], Carsten Sievers[1], Derrick W. Flick[2], Brien A. Stears[2], and John P. Henley[3]

[1]School of Chemical & Biomolecular Engineering, Georgia Institute of technology, Atlanta, Georgia 30332; [2]The Dow Chemical Company, Freeport, Texas 77541; [3]The Dow Chemical Company, Midland, Michigan 48674

Today, the world's demand for renewable or alternative resources for chemical and energy production is a key factor in the renewed exploration of pyrolysis and gasification of low cost carbon feedstocks such as biomass, low rank coal, and Municipal Solid Waste (MSW). Municipal Solid Waste (MSW), the everyday trash generated in households, is widely available at low cost in many locations. The biggest challenge with processing MSW into useful chemical or fuels form is its diverse nature which varies with location, time, and season. To obtain a representative MSW composition, data were collected from US EPA 2012 report (1). The major constituents of MSW were reported to be plastics (12.7%), paper and paperboard (27.4%), and biomass which includes food waste (14.5%), yard trimmings (13.5%) and wood (6.3%). However, a part of this generated MSW is recycled and/or used in heat recovery. The rest (53.8%) is either discarded in landfills or incinerated. Our goal is to study the sequential pyrolysis and gasification reactions of discarded MSW for the production of syngas.

Synthetic MSW was configured to mimic the composition in which it is likely to be procured. The plastics fraction consisted of Polyethylene (LDPE and HDPE), Polypropylene (PP), Polyethylene terephthalate (PET) and Polystyrene (PS), while the paper fraction consisted of newspaper, office paper, tissue paper and cardboard. The biomass fraction was a combination of food waste, grass and wood. Experiments with each of these components were carried out in a Thermogravimetric Analyzer (TGA) separately to understand the differences and similarities between their reactivity at different process conditions. Pyrolysis studies were carried out at temperatures between 30 – 1000 °C and at atmospheric pressure. Final temperature, heating rate and equivalence ratio (ER is defined as the ratio of air used / air needed for stoichiometric combustion) were varied to optimize pyrolysis and gasification conditions. DSC studies were carried out to understand the heat effects in the process.

In addition to the effect of process parameters, the effect of catalytic species present in various fractions of MSW will be discussed. It was observed that all the plastics, except PET, volatilize and gasify completely without leaving any char while paper and food waste always left a residue. The char generated by these fractions further undergoes gasification reactions. Further studies to investigate the combined effect of these individual components will be presented.

1. http://www.epa.gov/osw/nonhaz/municipal/pubs/2012_msw_fs.pdf

Rice husk as solid fuel for annular dual-tube moving bed reactor in chemical looping combustion

Young Ku[1], Hsin-Hsien Tsai[1], Hsuan-Chih Wu[1], Yu-Lin Kuo[2], Yao-Hsuan Tseng[1]

[1] Department of Chemical Engineering, National Taiwan University of Science and Technology, Taipei city, Taiwan

[2] Department of Mechanical Engineering, National Taiwan University of Science and Technology, Taipei city, Taiwan

Abstract

Rice husk as solid fuel for chemical looping combustion (CLC) with Fe_2O_3/Al_2O_3 oxygen carriers was investigated by an annular dual-tube moving bed reactor (ADMBR). CO and H_2 were the fuel gases generated by rice husk gasification in the ADMBR prior to combustion with oxygen carriers at 900 °C. 100% of CO_2 yield of fuel gases was achieved as oxygen carrier flow rate of rice husk reached 3.17. The fuel conversions of rice husk combustion were reached 92.56, for experiments conducted with oxygen carrier flow rate above 2.27 g/min higher than that gasification. The oxygen carriers that moving out of the ADMBR were composed of mainly $FeAl_2O_4$, characterized by X-ray diffraction (XRD) analysis. The formation of $FeAl_2O_4$ indicated that further utilization of oxygen in iron-based oxygen carriers can be achieved by moving bed operation. Chemical looping combustion is suggested to be an alternative treatment technology with high conversion for rice husk combustion.

Key words: chemical looping combustion (CLC); rich husk combustion; Fe_2O_3/Al_2O_3 composite oxygen carrier; moving bed reactor.

Oil Sorbents from Plastic Waste

Presenter

Junaid Saleem

Authors

Junaid Saleem, Alireza Bazargan, John Barford, and Gordon McKay

AIChE Annual Meeting 2014

Atlanta, USA, November 16-21, 2014

香港科技大學
THE HONG KONG UNIVERSITY OF SCIENCE AND TECHNOLOGY

Hong Kong University of Science & Technology

2

Outline of Presentation

- 1.0 Introduction
- 2.0 Synthesis of Oil Sorbent
- 3.0 Characterization
- 4.0 Results
- 5.0 Conclusion

3

1.0 Introduction - Plastic Polymer Waste

- Thermoplastic is a major constituent of municipal solid waste (MSW)
- 5-15% by weight
- 20-30% by volume
- Plastic Pollution Effect:
  - Land
  - Ocean
  - Animals
  - Humans

http://www.plasticfreeocean.org/node/18

4

1.0 Introduction - Traditional Methods of Overcoming Plastic Wastes

- Landfilling
  - Eventually will be phased out due the scarcity of space and its ever increasing cost
- Incineration
  - Emission of Toxic Fumes
  - Generation of Fly-Ash that requires disposal

1.0 Introduction
Up-cycling/Upgrading of Plastic Wastes

- Objective is to make a value added product that:
  - Reduces the amount of plastic waste pollution
  - Reduces water pollution by oil spill clean-up remedy

Oil Spill Problems

1.0 Introduction - Oil Spill Hazards

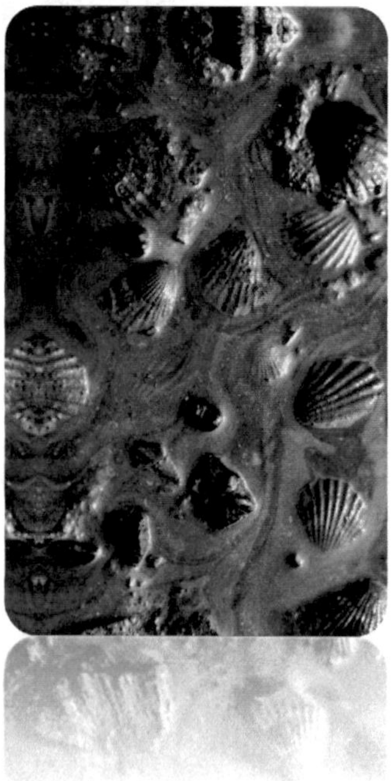

- Environmental effects
  - Scent of animals fade due to strong smell of oil
  - Impair birds' ability to fly
  - Ingestion damages kidney and liver function
  - Changes in hormonal balance
  - Reducing insulating effect of fur and plumage
  - Growth of oil-consuming bacteria
- Impact on humans
  - Explosion and fire hazard
  - Drinking water contamination, respiratory diseases, etc.
  - Damage to tourism

1.0 Introduction - Methods of Combating Oil Spills

- Methods of combating oil spills:
  - Bio remediating
  - Controlled burning
  - Dispersing
  - Dredging
  - Skimming
  - Vacuuming
  - Shovelling
  - Solidifying/absorbing/adsorbing
  - Watching and waiting!

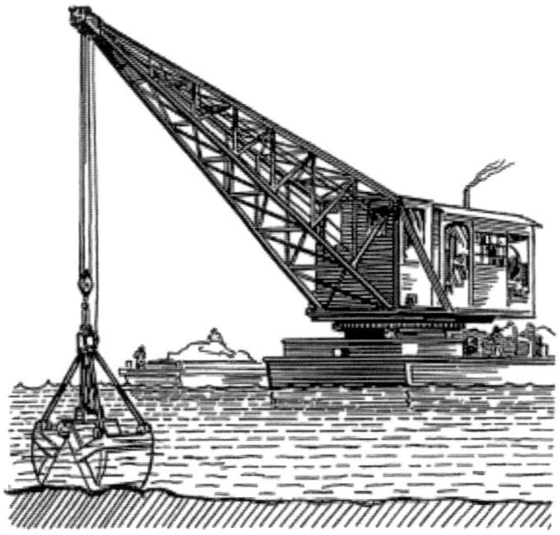

- Factors for choosing clean up method(s): temperature, weather condition, type of liquid, amount of liquid, type of shoreline etc.

1.0 Introduction - Classification of Plastic Wastes

Polymer	% Plastic Waste
LDPE/LLDPE	20.6
PP	20
HDPE	17.4
PET	11.7
PS	10.9
PVC	10.9
PU	2.7
Others	9.8

HDPE bottles were used to prepare Oil-Sorbent, supplied by Nestle Hong Kong

Bazargan et. al. 2013, and Baroulaki et. al. 2006

Part 2.0

OIL SORBENT FOR OIL SPILL REMEDY

2.0 Quick and High uptake

- First few hours are crucial to control an oil spill
- Studies on Persian Gulf and Singapore Coastal Water:
  - Thickness of an oil spill slick is a strong function of time
  - Oil slick thickness rapidly decreases during first 3 hours
- The need exists to prepare a quick and high oil uptake sorbent

R. Ghiassi et. al. 2012 and Chao et. al. 2003

2.0 Properties of Polymeric Oil Sorbent

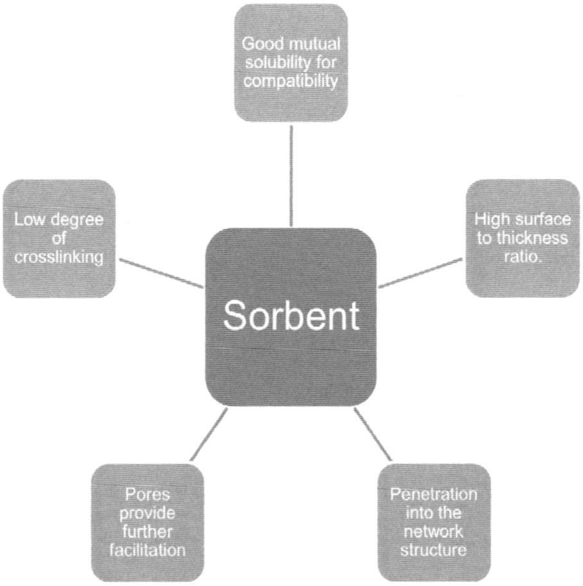

Ref: [US4039489]

2.0 Sorbent Preparation Methodology

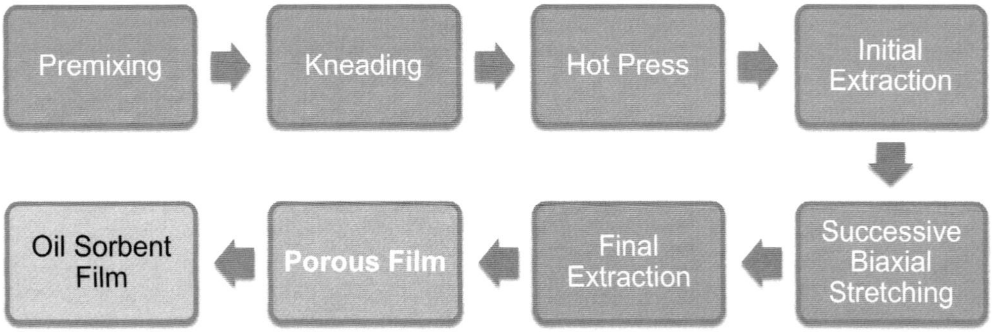

- Raw materials: Waste HDPE bottle (50%), UHMWPE (50%), Paraffin Wax for swelling and for creating pores.
- 50 % UHMWPE virgin powder (3.5 Million molar mass) used to give sorbent sufficient strength
- Waste HDPE bottles were supplied by Nestle Hong Kong Ltd.

2.0 Oil Sorption Uptake Test

- HD-UH (50%HDPE and 50% UHMWPE) porous film of 3x3 cm^2 was placed on the surface of marine diesel oil
- Contact time of sorbent in sorbate are: 0.5, 1, 2, and 5 mins.
- Sorbent was allowed to hang for different dripping durations: 2, 5, 10, 15, 30, 45, and 60 mins.
- Final wt. of sorbent was measured after each particular dripping

2.0 Process of Oil Sorption

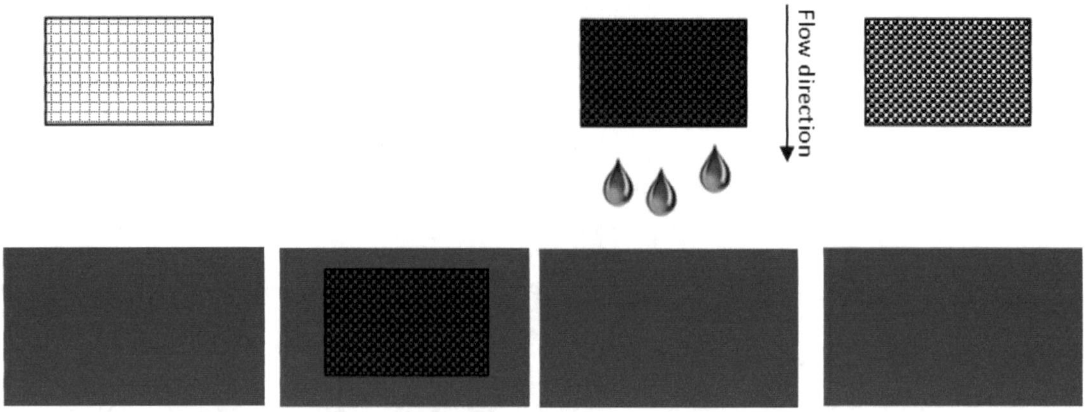

Step 1: the sorbent is prepared and weighed	Step 2: the sorbent is submerged in the sorbate	Step 3: the sorbent is removed and drained	Step 4: equilibrium is reached after dripping

Part 3

CHARACTERIZATION

3.0 Differential Scanning Calorimetry - DSC

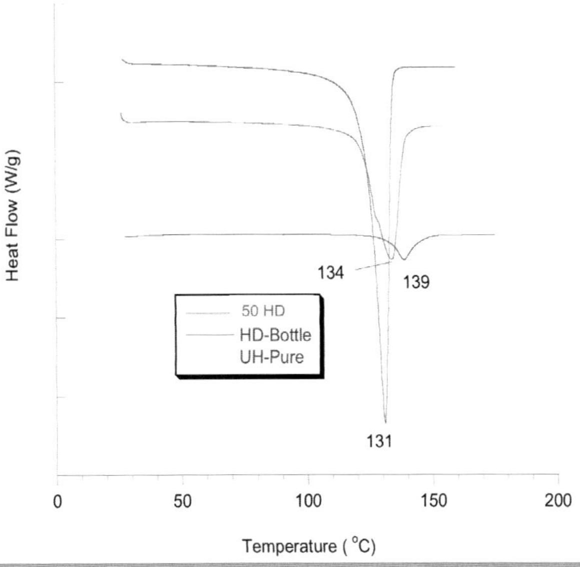

- DSC showing melting peaks of samples
- 50HD gives one melting peak at 134 °C which is in between UHMWPE (139 °C) and HDPE waste bottle (131 °C)

3.0 Scanning Electron Microscopy - SEM

- 50% Waste HDPE + 50% UHMWPE

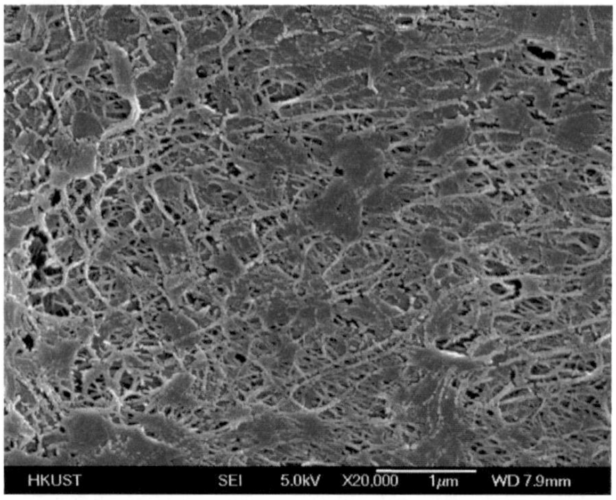

SEM image showing porous structure (70%)

4.0 Results - Sorption Capacity

- Sorption Capacities Using 4 Oils

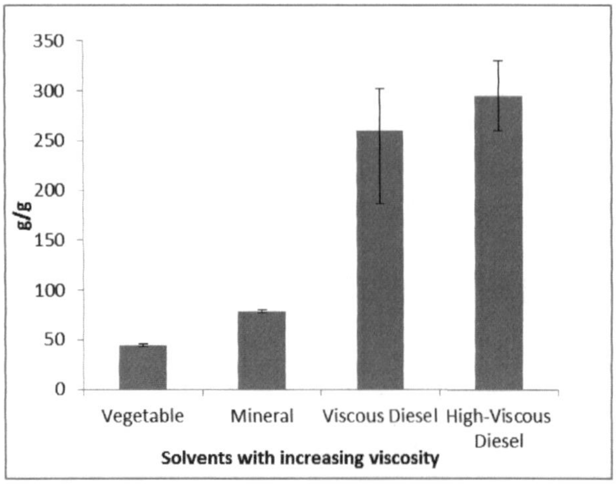

Solvent	Viscosity* Pa.s	Density g/cm³
Corn oil	0.09	0.853
Mineral oil	0.21	0.831
Marine diesel oil	7.13	0.991
Highly viscous diesel oil	13.82	0.991

* At Shear rate of 1/sec and at 23°C

4.0 High Capacity: A Combination of Absorption, Adsorption and Adhesion

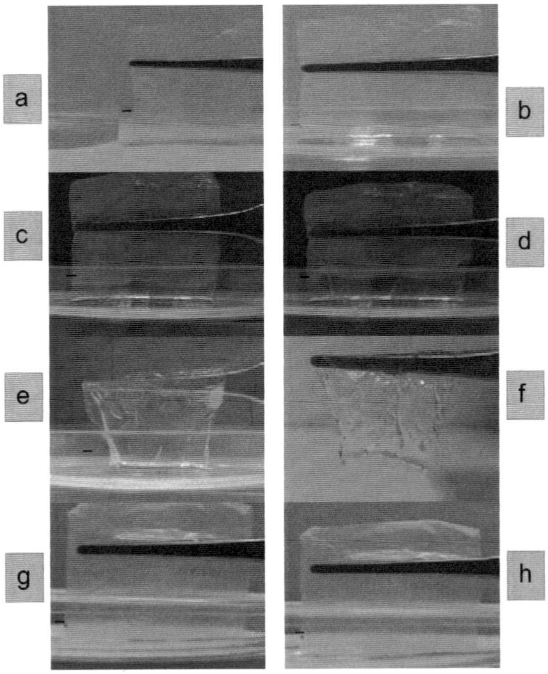

Figure 1:

(a) Sorbent film (size 3x3 cm2) before being immersed in corn oil bath,
(b) After immersing,
(c) After 30 mins.,
(d) After 60 mins.,
(e) After 12 hrs.,
(f) Adhesion of oil at the edge of film,
(g) Film immersed in water, and
(h) Film in water after 12 hrs.

Contact Angle: 115°

21

4.0 Dripping Kinetics of the 50% HD Sorbent using High Viscous Diesel Oil (marine diesel)

Oil Sorbed (g/g) versus Time (min)

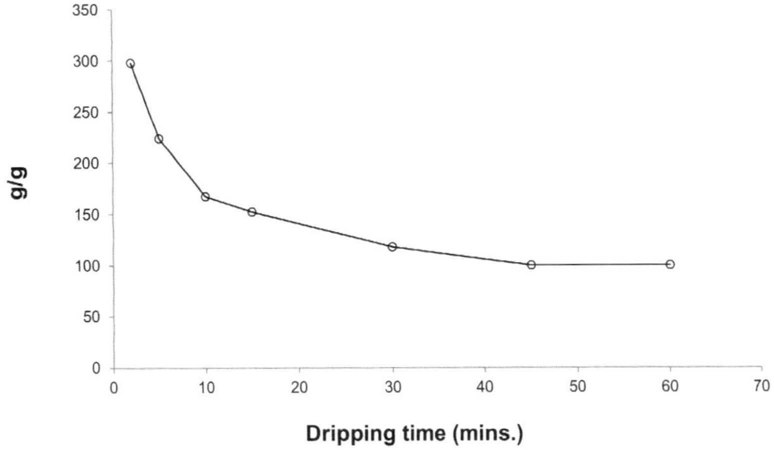

22

4.0 Saturation Contact Time Study (Sorbate- High viscous diesel oil)

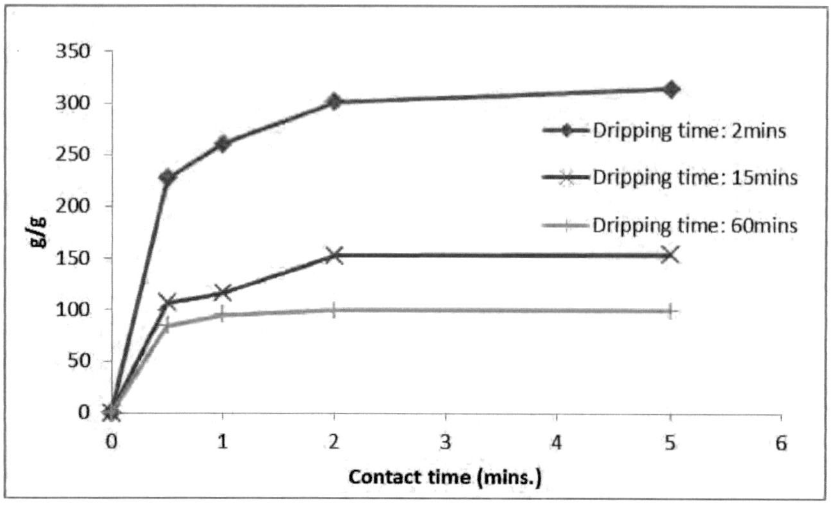

4.0 Mechanical Properties

Sample	Modulus MPa	Tensile Strength MPa
U-HD-50	342	80
PE-HMW	1332	171
Kapton®	2731	177
A4 Paper	1343	34
Aluminum foil	664	69
Parafilm	2.5	2.06

4.0 Fitting of Experimental Data

Four "sorption" kinetic models were tested: Pseudo First order, Pseudo Second order, Elovich and Ritchie Kinetic Models

The pseudo second- order model proposed by Ho and McKay (Co-author of current work) was found to be the best fit

$$\frac{1}{Q_m - Q_t} = \frac{1}{Q_m} + kt$$

where Q_m = saturated sorption capacity
Q_t = sorption capacity (g/g) at time t
k = sorption constant

25

4.0 Experimental Data fits Pseudo Second Order Model

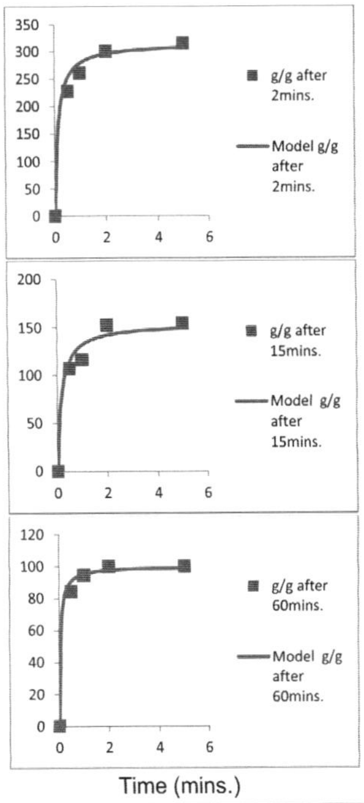

26

5.0 Conclusions

- Prepared value added oil-sorbent from plastic waste
- Successfully prepared oil sorbent with uptake capacity of 44 to 300 g/g depending on oil type
- Oil uptake saturation reached from 0.5 to 2 minutes depending on oil type
- Complete dripping profile was performed
- Experimental results fit pseudo second order kinetics sorption model

THANK YOU

386125 Waste to Energy: Reaction Engineering Routes to Waste Gasification for Sustainable Living Environments

Abstract:

This research initiative focuses on advancing a low-to-mid temperature catalytic gasification process as a potential resource utilization alternative. The main goal of this project is the characterization of reaction pathways as waste management and sustainable energy alternatives for waste management applications. The project is motivated by the need to develop low-cost technologies as environmentally-friendly alternatives for converting waste materials into high-value products. Indeed, as landfills are reaching capacity, waste management is becoming an increasingly important issue for large residential areas. The completion of the study is anticipated to lay out the foundation for scaling-up the technology to a continuous process.

385315 Pyrolysis of Plastic Wastes

Abstract:

Recycling of plastics is both challenging and non-economic due to the presence of impurities and the energy intensive recovery process. Pyrolysis, as thermochemical conversion of waste to recovery lower molecular weight hydrocarbon products, is an attractive process which simultaneously reduces emissions by locking carbon (char prodcution).

Experiments are reported where plastic wastes are pyrolysed by employing a laboratory scale batch fixed bed reactor. A quantitative analysis of the gas, liquid, wax and solid products was conducted to study the process. TGA /DSC measurements were undertaken to investigate the volatile yields, char-like yields, and the fixed carbon produced. Subsequently, the experimental results were used to validate kinetic lumping models developed to describe the primary and secondary products of the pyrolysis process. The kinetic parameters were estimated as function of time at three different operating temperatures and the model was fully characterised by identifying the kinetic rate constants and the activation energies of the primary and secondary pyrolysis products.

ANALYSIS OF THE PRODUCTION OF BIOFERTILIZERS FROM COLOMBIAN LIGNOCELLULOSIC FEEDSTOCKS

Valentina Hernández[1], Laura V. Daza[1], Miguel A. Rojas[1], Eulogio Castro[2], Carlos A. Cardona[1]

[1]Instituto de Biotecnología y Agroindustria, Universidad Nacional de Colombia Sede Manizales

[2]Dept. Chemical, Environmental and Materials Engineering. Agrifood Campus of International Excellence (ceiA3), University of Jaén, 23071 Jaén, Spain

14AIChE
Annual Meeting, Atlanta, GA

Outline

Research Group in Chemical, Catalytic and Biotechnologic Processes – UNAL
Research Group in Chemical and Environmental Engineering - UJA

Economic and environmental comparison of Colombian lignocellulosic materials for
- biofertilizers (anaerobic digestion)
- electricity (gasification)

1. Introduction
2. Methodology
3. Results and Discussion
4. Conclusions

2

Research Group in Chemical, Catalytic and Biotechnologic Processes – UNAL
Research Group in Chemical and Environmental Engineering - UJA

1. Introduction

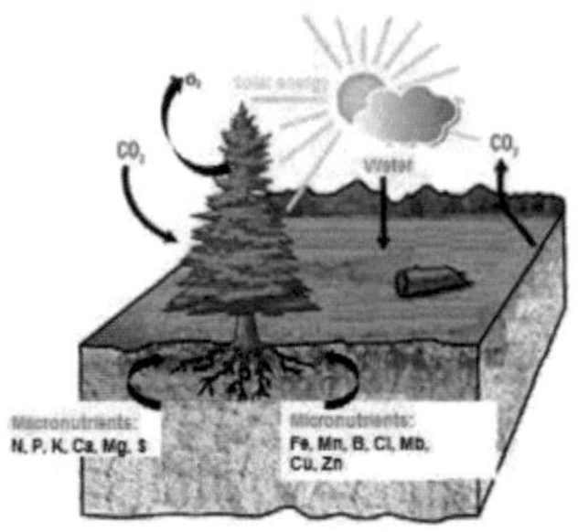

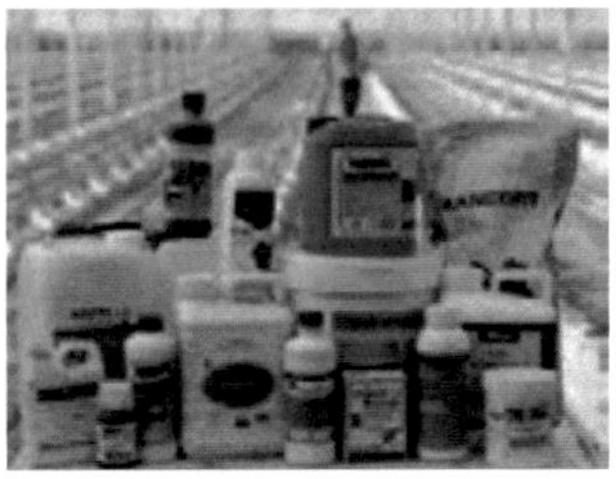

By 2050 → 9 billion people
At least 50% of the food demand should be supplied by already cultivated soils

3

Research Group in Chemical, Catalytic and Biotechnologic Processes – UNAL
Research Group in Chemical and Environmental Engineering - UJA

1. Introduction

- Soil deployment
- GHG
- Contamination of underground water
- Petroleoum dependency
- Extraction costs

→ 7 – 33 %

In Colombia fertilizers represent 7 to 33 % of the production cost

4

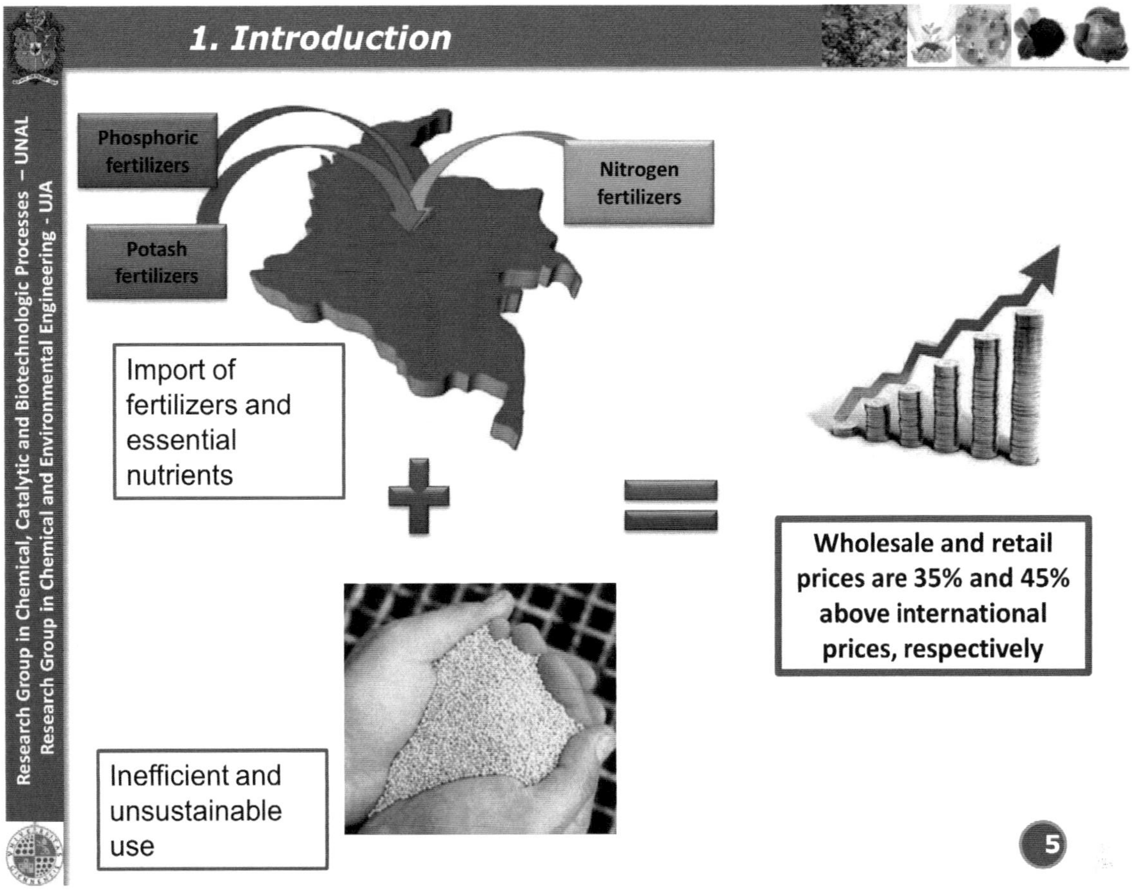
1. Introduction
Research Group in Chemical, Catalytic and Biotechnologic Processes – UNAL
Research Group in Chemical and Environmental Engineering - UJA
Phosphoric fertilizers
Potash fertilizers
Nitrogen fertilizers
Import of fertilizers and essential nutrients
Inefficient and unsustainable use
Wholesale and retail prices are 35% and 45% above international prices, respectively
5

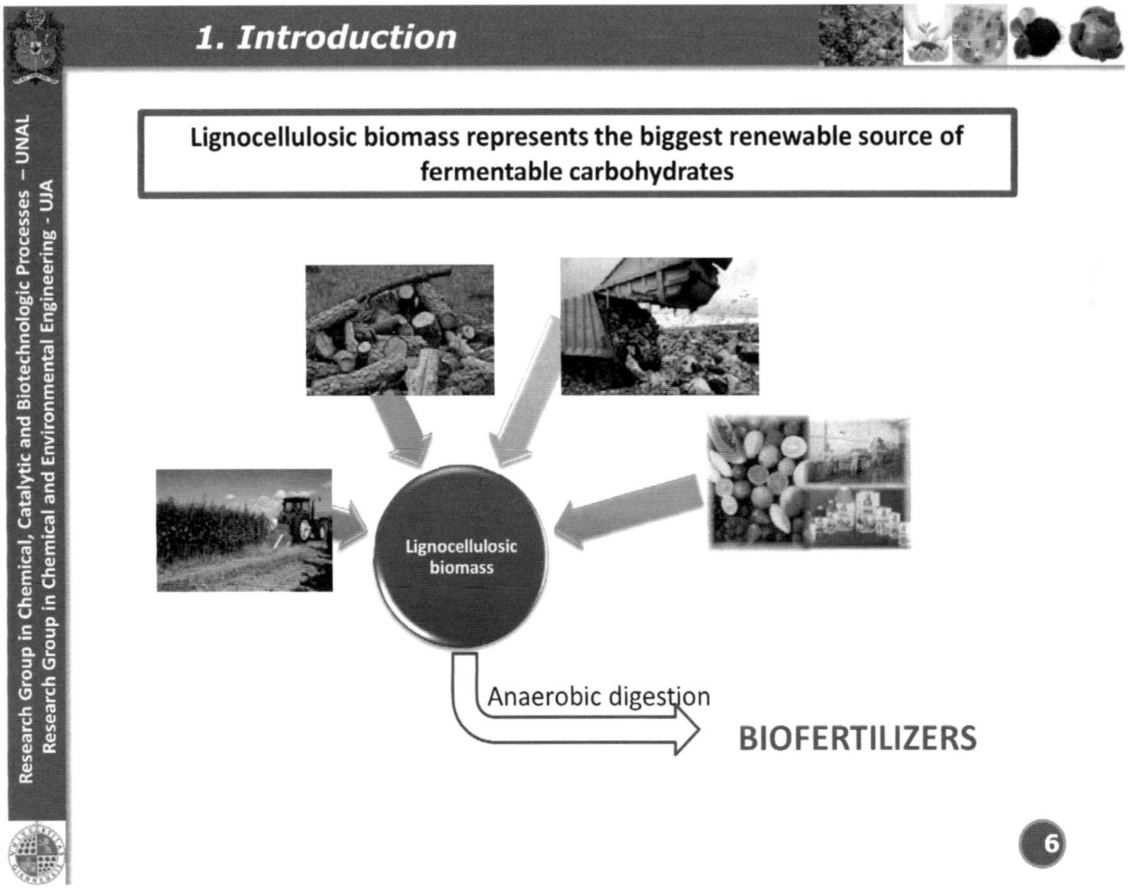
1. Introduction
Research Group in Chemical, Catalytic and Biotechnologic Processes – UNAL
Research Group in Chemical and Environmental Engineering - UJA
Lignocellulosic biomass represents the biggest renewable source of fermentable carbohydrates
Lignocellulosic biomass
Anaerobic digestion
BIOFERTILIZERS
6

1. Introduction: Anaerobic digestion

Research Group in Chemical, Catalytic and Biotechnologic Processes – UNAL
Research Group in Chemical and Environmental Engineering - UJA

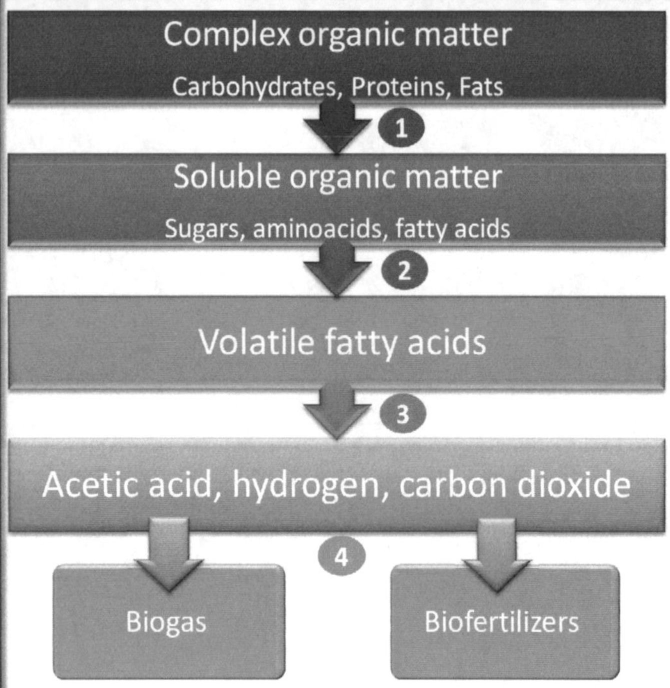

① Hydrolysis
② Fermentation
③ Acetogenesis
④ Methanogenesis

Approximatelly 90% of the feedstock is converted into biofertilizers

Nutrients from the feedstock are recovered in the final solid product

7

1. Introduction

Research Group in Chemical, Catalytic and Biotechnologic Processes – UNAL
Research Group in Chemical and Environmental Engineering - UJA

OBJECTIVE: comparison of lignocellulosic materials for

Fertilizer production by pretreatment+anaerobic digestion

Electricity production by gasification (base case)

Lignocellulosic materials
- Largely available in Colombia
- Residual biomass
- No alternative uses

ORANGE PEELS
COFFEE HUSKS
PLANTAIN EMPTY FRUIT BUNCHES

Economic comparison
- ASPEN Plus/Aspen Plus Economic Analyzer

Environmental assessment
- Waste Algorithm Reduction
- GHG assessment

8

2. Methodology

Simulation of the anaerobic digestion

- Mass & Energy balances → Aspen Plus® v7.3 → raw materials, consumables, utilities and energy requirements.
- Energy consumption → Thermal energy required in heat exchangers and reboilers

Economic assessment

- Capital & operating costs → Aspen Economic Analyzer v7.3 → Production costs
  - Colombian case, 10 years
  - Linear depreciation

Environmental assessment

WAR Waste Algorithm Reduction, from Environmental Protection Agency –EPA. Potential Impact Factor

- HTPI, Human Toxicity by Ingestion;
- HTPE, Human Toxicity by Dermal Exposition or Inhalation;
- ATP, Aquatic Toxicity Potential;
- GWP, Global Warming Potential;
- ODP, Ozone Depletion Potential;
- PCOP, Photochemical Oxidation Potential;
- AP, Acidification Potential

9

Research Group in Chemical, Catalytic and Biotechnologic Processes – UNAL
Research Group in Chemical and Environmental Engineering - UJA

2. Methodology: Anaerobic digestion

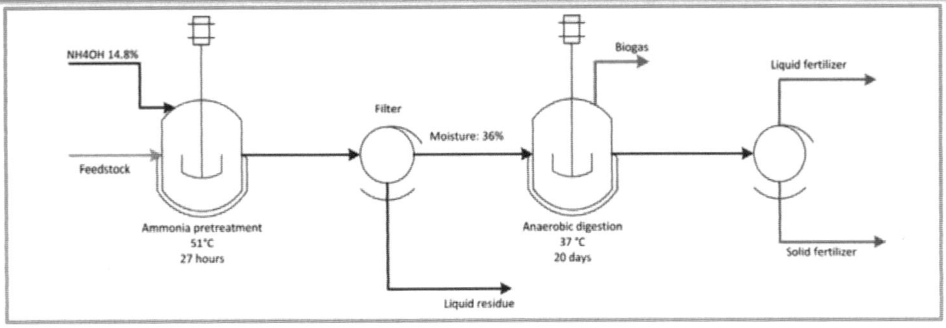

Anaerobic digestion: Kinetic model

$$-\frac{dS_{VSS}}{dt} = k_1 S_{VSS}$$

$$\frac{dS_{VDS}}{dt} = k_1 S_{VSS} - \frac{k_2 S_{VDS}}{k_3 + S_{VDS}}$$

$$\frac{dS_{TVA}}{dt} = \frac{k_4 S_{VDS}}{k_5 + S_{VDS}} - \frac{k_6 S_{TVA}}{k_7 + S_{TVA}}$$

$$\frac{dS_M}{dt} = \frac{k_6 S_{TVA}}{k_7 + S_{TVA}}$$

k_1=0.054 day-1	k_5=10.2 g/l
k_2=4.2 g/l	k_6=4.3 g/l
k_3=9.8 g/l	k_7=3.1
k_4=3.6 g/l	

VSS	Volatile suspended solids
VDS	Volatile dissolved solids
TVA	Total volatile acids
M	Methane

Borja et al., Process Biochemistry. 2005;40:1841-7.

10

Research Group in Chemical, Catalytic and Biotechnologic Processes – UNAL
Research Group in Chemical and Environmental Engineering - UJA

2. Methodology: Gasification

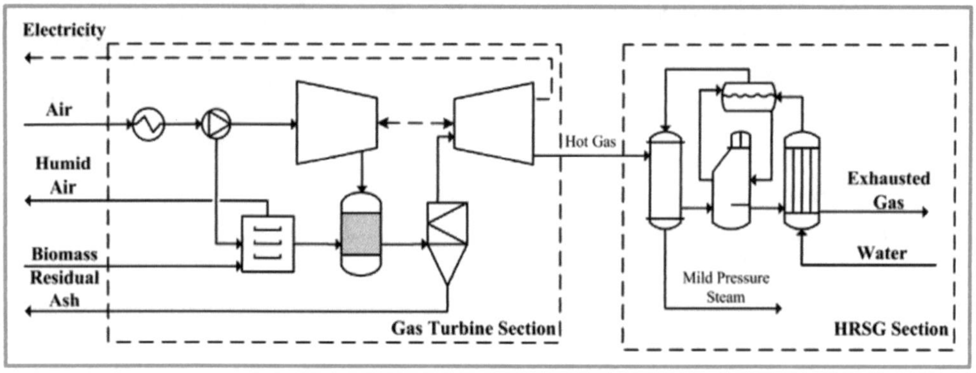

Biomass Integrated Gasification Combined Cycle (BIGCC):

Biomass dryer
Gasification chamber
Gas turbine and heat steam recovery generator (HRSG)

11

Research Group in Chemical, Catalytic and Biotechnologic Processes – UNAL
Research Group in Chemical and Environmental Engineering - UJA

2. Methodology: GHG assessment

Goal and scope

- To establish the general purpose of the assessment
- To describe the system
- To establish the system boundaries
- To select the functional unit
- To select the allocation method

Inventory analysis

- Based on mass and energy global inputs and outputs

Impact assessment

- To carried out the GHG balance based on the individual emission factors

12

3. Results and discussion

Research Group in Chemical, Catalytic and Biotechnologic Processes – UNAL
Research Group in Chemical and Environmental Engineering - UJA

Raw material characterization

Parameter (dry basis)	Raw material		
	Orange peel	Coffee husk	Plantain empty fruit bunches
Cellulose	15.42	34.38	18.06
Hemicellulose	17.23	25.55	4.79
Lignin	11.58	24.38	24.43
Extractives	51.81	11.51	49.35
Ash	3.18	0.65	2.43
O-acilo group	0.79	3.52	0.92
NUTRIENTS			
Nitrogen (%)	1.109	0.933	0.3113
Phosphorous (ppb)	2.334	1.043	0.9429
Potassium (ppb)	998.8	2,334.5	2,393.1
Moisture	8.46	10.50	9.15

13

3. Results and discussion

Research Group in Chemical, Catalytic and Biotechnologic Processes – UNAL
Research Group in Chemical and Environmental Engineering - UJA

Economic assessment

PRODUCTION CAPACITIES			
Stream	Orange peel	Coffee husk	Plantain empty fruit bunches
ANAEROBIC DIGESTION			
Feedstock (kg/day)	1,000	1,000	1,000
Liquid fertilizer (kg/day)	13.42	25.26	14.92
Solid fertilizer (kg/day)	4.75	8.95	5.28
Biogas (m^3/day)	18.39	34.63	12.59
BASE CASE (GASIFICATION)			
Electricity (MW-h)	0.407	0.345	0.309

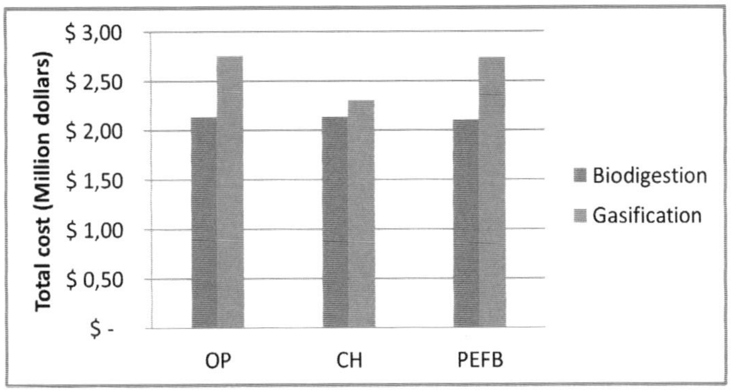

14

Research Group in Chemical, Catalytic and Biotechnologic Processes – UNAL
Research Group in Chemical and Environmental Engineering - UJA

3. Results and discussion

Economic assessment

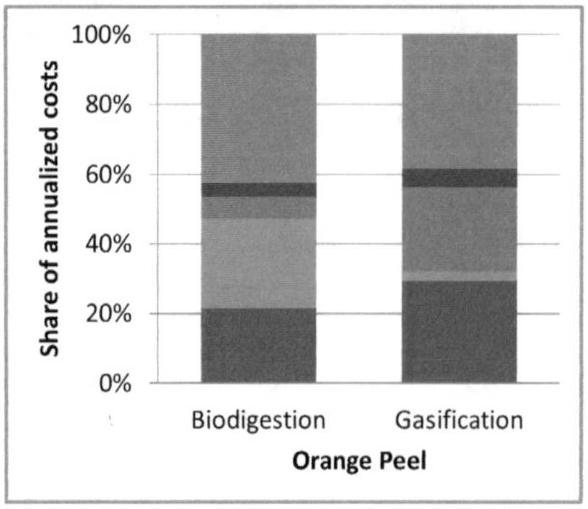

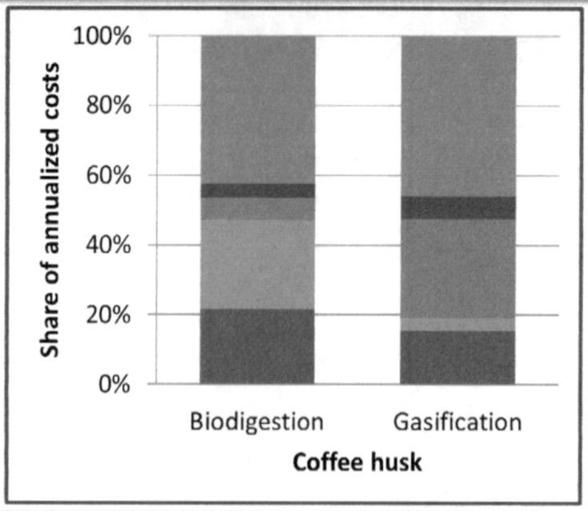

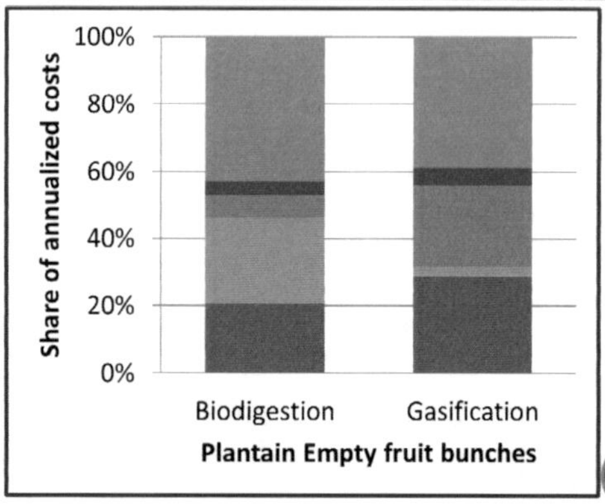

- **Operating costs**
- Operating Labor and Maintenance
- Utilities
- Raw material
- Depreciation

15

Research Group in Chemical, Catalytic and Biotechnologic Processes – UNAL
Research Group in Chemical and Environmental Engineering - UJA

3. Results and discussion

Environmental assessment

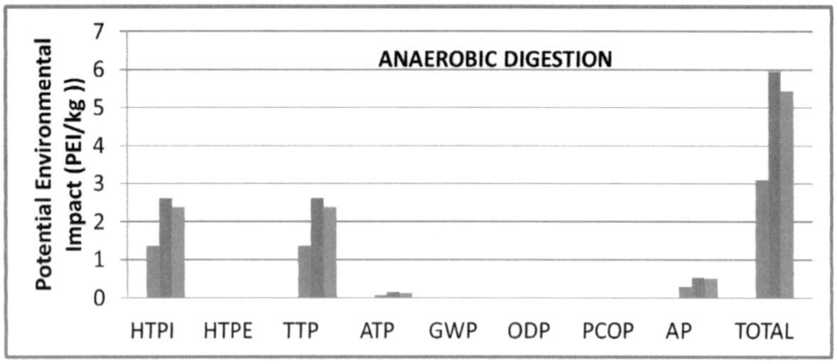

- CH
- OP
- PEFB

HTPI, Human Toxicity Ingestion
HTPE, Human Toxicity by Dermal Exposition or Inhalation
TTP, Terrestrial Toxicity Potential
ATP, Aquatic Toxicity Potential
GWP, Global Warming Potential
ODP, Ozone Depletion Potential
PCOP, Photochemical Oxidation Potential;
AP, Acidification Potential

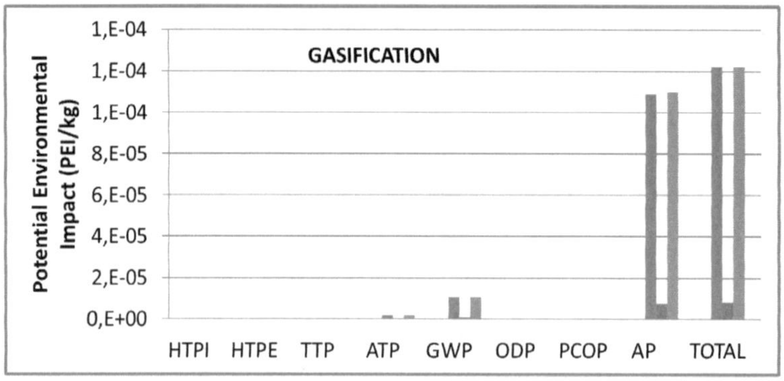

16

3. Results and discussion

GHG assessment

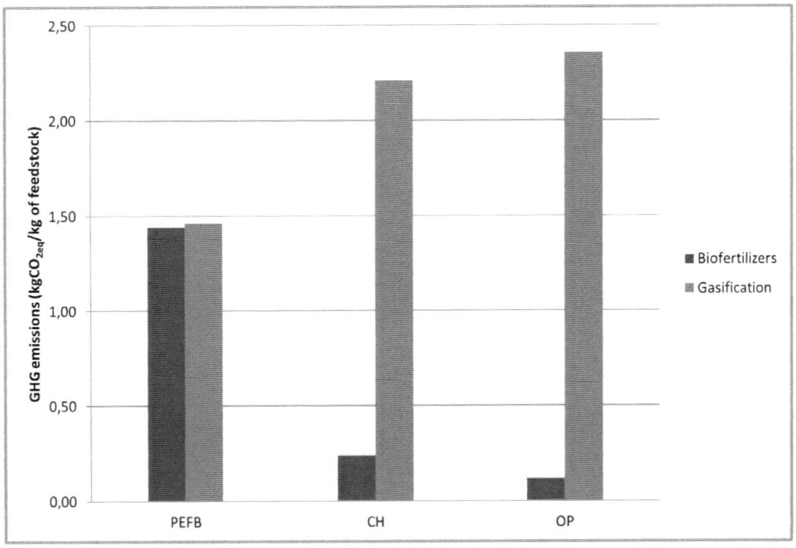

Research Group in Chemical, Catalytic and Biotechnologic Processes – UNAL
Research Group in Chemical and Environmental Engineering - UJA

17

4. Conclusions

Based on ASPEN Plus® and WAR/GHG analyses

- ❑ Orange peel, coffee husks and plantain empty fruit bunches are better used for biofertilizer production compared to gasification.
- ❑ Annualized costs for the biofertilizer production are 5-21% lower than those for gasification
- ❑ From the environmental point of view, GHG emissions associated to fertilizer production are lower than those from gasification

Research Group in Chemical, Catalytic and Biotechnologic Processes – UNAL
Research Group in Chemical and Environmental Engineering - UJA

18

Research Group in Chemical, Catalytic and Biotechnologic Processes – UNAL
Research Group in Chemical and Environmental Engineering - UJA

Acknowledgements

- Universidad Nacional de Colombia at Manizales
- Investigation and extension direction, from the Engineering and Architecture Faculty from Universidad de Colombia Sede Manizales
- Design of Sustainable Biorefineries (code 157) and Chemical, catalytic and biotechnological Processes (code 16081) projects from the Investigation Direction at Manizales (DIMA)
- Administrative Department of Science, Technology and Innovation (COLCIENCIAS)
- The Agrifood Excellence Campus (ceiA3) in University of Jaén
- The Agrifood International Doctorate School (eidA3), Spain
- Junta de Andalucía, Spain (Proyecto de Excelencia AGR-6103)

19

Research Group in Chemical, Catalytic and Biotechnologic Processes – UNAL
Research Group in Chemical and Environmental Engineering - UJA

Thanks for your kind attention

20

References

- Borja, R. et al. (2005). "Kinetic modelling of the hydrolysis, acidogenic and methanogenic steps in the anaerobic digestion of two-phase olive pomace (TPOP)." Process Biochemistry **40**(5): 1841-1847.
- Consejo Nacional de Política Económica y Social (2009). Política Nacional para la Nacionalización del Componente de Costos de Producción Asociado a los Fertilizantes en el Sector Agropecuario. Documento Conpes 3577.
- Chirinda, N., J. E. Olesen, et al. (2010). "Soil properties, crop production and greenhouse gas emissions from organic and inorganic fertilizer-based arable cropping systems." Agriculture, Ecosystems & Environment **139**(4): 584-594.
- Food and Agriculture Organization of the United Nations (2008). "Current World Fertilizer Trends and Outlook to 2011/12."
- Food and Agriculture Organization of the United Nations (2011). "Current World Fertilizer Trends and Outlook to 2015."
- Food and Agriculture Organization of the United Nations (2012). "World Agriculture Towards 2030/2050: The 2012 revision."
- Food and Agriculture Organization of the United Nations and International Fertilizer Agency (2002). "Los Fertilizantes y su Uso." **Cuarta edición**.
- Li, Y., U. Merrettig-Bruns, et al. (2014). "Optimization of ammonia pretreatment of wheat straw for biogas production." Journal of Chemical Technology & Biotechnology: n/a-n/a.
- Rasool, R., S. S. Kukal, et al. (2007). "Soil physical fertility and crop performance as affected by long term application of FYM and inorganic fertilizers in rice–wheat system." Soil and Tillage Research **96**(1–2): 64-72.
- Suárez, A. (2013) Colombia, Campeón Mundial en Precio de Fertilizantes. El Espectador

Research Group in Chemical, Catalytic and Biotechnologic Processes – UNAL
Research Group in Chemical and Environmental Engineering - UJA

21

Production of Renewable Aromatic Hydrocarbons from Waste Coffee Cups over Zeolites via Catalytic Pyrolysis

Honghong Shi, Rico Silbermann, Aqsha Aqsha and Nader Mahinpey*
Energy and Environment Research Group (EERG)
Department of Chemical and Petroleum Engineering
University of Calgary
Calgary, Alberta, Canada
T2N 1N4

Aromatic hydrocarbons; especially benzene, toluene, and xylene (BTX); are very important intermediates for a wide variety of downstream chemicals. In this study, waste paper coffee cups were used as feedstock to produce aromatics hydrocarbons through catalytic pyrolysis. Waste paper coffee cups has the potential as energy feedstock due to the fact that around 1.6 billion [1] paper coffee cups are abandoned everyday worldwide. From our proximate analysis, paper coffee cup has high volatile matter content of 91.7%, while the ash content is only 1.09%. Another advantage of coffee cups is their low lignin content due to the necessary chemical removal procedure in papermaking process. Lignin has been proved to form coke on the surface of catalyst rather than form liquid products [2].

The gallium promoted ZSM zeolites were carefully prepared and tested in a bench scale vertical fixed bed reactor along with an online Micro GC as well as an offline GC-MS to analyze the outlet gas and liquid compositions. Zinc was also investigated as an alternative metal candidate to enhance the performance of zeolite catalyst. Comparative studies of different catalyst loading patterns and different reaction temperatures (from 400 °C to 550 °C) were conducted. The catalyst was subjected to 10 reaction-regeneration cycles, in order to assess its stability. After each reaction, the catalyst was purged and regenerated in air for approximately 3 hours. Various characterizations of fresh, spent and regenerated catalysts were carried out using X-Ray Diffraction (XRD), Scanning Electron Microscopy (SEM), Transmission Electron Microscopy (TEM), Fourier Transform Infra-Red Spectroscopy (FTIR) and Ammonia Temperature Programmed Desorption (TPD) to gain a clear insight into the catalyst properties. The results of this study suggest a renewable path way to produce aromatic hydrocarbons from wastes and a better understanding of the catalytic mechanism.

Keywords: aromatics, catalytic pyrolysis, zeolite, waste coffee cups

1. International Coffee Organization. Available from: http://www.ico.org/show_faq.asp?show=35.
2. Jae, J., et al., *Depolymerization of lignocellulosic biomass to fuel precursors: maximizing carbon efficiency by combining hydrolysis with pyrolysis.* Energy & Environmental Science, 2010. **3**(3): p. 358-365.

*Author for correspondence:

Dr. Nader Mahinpey
Dept. of Chemical and Petroleum Engineering
Schulich School of Engineering
The University of Calgary
2500 University Drive NW
Calgary, AB T2N 1N4 Canada
Phone: (403) 210-6503
Fax: (403) 284-4852
E-mail: nmahinpe@ucalgary.ca

Please for all correspondence on this article, kindly contact the principal author (Dr. Nader Mahinpey) as stated above. Thank you.

Integrated waste treatment management in lignocellulosic biorefineries

Aikaterini D. Mountraki, Marinella L. Tsakalova, Antonis C. Kokossis
School of Chemical Engineering, National Technical University of Athens, Greece

Abstract

This paper outlines a synthesis approach for the integrated management of waste in multi-product biorefineries. Waste treatment is not restricted to handling, but it is considered as a process which can offer valuable products. An integrated approach can reduce treatment costs and add value to those by-products that are impossible to include in the value chain.

The work introduces a systematic methodology based on synthesis and optimization technologies. A syperstructure approach is set up to screen the waste management technologies and to select treatment policies either as single or hybrid solutions. The systematization is based on a preliminary stream classification that is used to allocate by-product streams to potential treatment units. The number, type, capacity and the particular waste treatment technology are addressed as degrees of freedom to optimize. The allocation of streams to technologies and processing units yields a synthesis model that is formulated using cost models for CAPEX and OPEX to trade against benefits produced by the use of waste (cogeneration, biogas production etc).

The synthesis model is optimized using discrete-continuous optimization techniques and can be extended with additional objectives that account LCA metrics. The method has been applied and illustrated in an organosolv biorefinery that produces sugars (C6 and C5) and lignin as intermediates. The optimization has been applied to assess the benefits in using the integrated approach, sensitivities associated with technology uncertainties and the impact of product portfolios in the analysis.

Keywords: Waste handling, Synthesis, Biorefinery, Lignocellulosic, Mixed Integer.

388761 Process Simulation of Integrated Liquid-State and Solid-State Anaerobic Digestion of Dairy Manure and Crop Residues

Abstract:

Dairy manure is the most common agricultural digester feedstock in the United States. The bioenergy from dairy manure is produced by producing biogas using a technique, liquid state anaerobic digestion (LAD). The major problem in LAD is disposal of large volume of effluent, which also produce odor problem if it is used to irrigate the farms. Crop residues can be used in solid state anaerobic digestion (SS-AD) to produce biogas. Advantages of SS-AD include smaller reactor capacity requirements, less energy used for heating, and no processing energy needed for stirring. Due to its lower water content, the digestate of SS-AD can be used as fertilizer and pelletized fuel, so it is much easier to handle than the effluent of LAD. Disadvantage of SS-AD systems are requirement of larger amounts of inoculum and longer retention time. By integrating LAD and SS-AD, effluent disposal problem of LAD and larger amounts of inoculum need of SS-AD is possible to resolve. The effluent from LAD is fed into SS-AD to wet the crop residues and add more inoculum for anaerobic digestion. The digestate comes from the SS-AD, which has no problem of odor issue can be used in the farm for fertilizing. In order to evaluate the cost and energy efficiency of this integrated AD system, process simulation model was developed using Super Pro design software. An accurate computer simulation process of the complete operation is developed to improve the operation of designed integrated anaerobic digestion system. The goal is to expand the dynamic anaerobic reactor model to improve its ability and simulate continuous liquid anaerobic reactor designs and continuously feed the SS-AD where crop residues and other solid waste are digested to increase the biogas production substantially. This process simulation model was developed by assuming 90% of the manure is collected from 500 head of cattle, manure is produced and collected constantly throughout the year, manure is managed as liquid and is free of bedding materials. For SS-AD, crop residues such as wheat straw and corn stover were considered for digestion. Monod kinetic model was used for anaerobic digestion reaction of organic matter to produce biogas in both reactors. The produced biogas was further cleared remove carbon dioxide and sulfur and converted into compressed natural gas or used for production of electricity. The capital and operating costs of producing clean biogas from the integrated LAD and SS-AD were presented.

389589 A Two-Stage Process for Converting Waste Plastics into Energy

Abstract:

In this work, a laboratory-scale two-stage reactor for converting waste plastics into energy was constructed. Our process converts waste plastics into synthesis gas (hydrogen and carbon monoxide) and methane that can be used for energy production. This innovative reactor technology employs a pyrolysis reactor as the first step, followed by a chemical looping gasification reactor as the second step. This two-stage process is more suitable than the traditional one-step gasification process, because solid residues associated with the waste, such as ash and tar, are contained in the pyrolysis unit. In addition, our reactor system utilizes metal oxides (Me_xO_y) in the gasification (second) step instead of traditional oxidizers such as oxygen or steam. This further saves the cost for the expensive air separation step to separate desired products from nitrogen.

Our experimental results using polyethylene as a surrogate plastic waste showed that the two-step pyrolysis-looping process significantly increased the yields of light gaseous products (hydrogen, water, carbon monoxide, carbon dioxide, and methane) compared to only pyrolysis. Polyethylene can be converted into light gaseous products in the presence of iron (III) oxide (Fe_2O_3) oxygen carriers. The total carbon-based mass yields of these products reached as high as 53%, and the total hydrogen-based mass yields reached as high as 86%. The product distributions of light gaseous products can be varied by certain controllable parameters (e.g., metal oxide loading, carrier gas flow rates, and gasification temperature), and key parameters affecting product yields were identified as gasification temperature and metal oxide loading and deactivation.

CO_2 can be utilized efficiently through a process of electrochemical conversion to valuable chemicals such as mono-ethylene glycol (MEG). This abstract reports part of the process engineering efforts undertaking at Liquid Light, Inc. in the development of an integrated commercial process that utilizes CO_2 as feedstock and produces fiber grade MEG at competitive cost advantages over best available technologies.

This presentation focuses on one of the most important process engineering challenges in the CO_2 to MEG technology: esterification and separation of oxalic acid ester in difficult chemical and physical systems from electrochemical cell operation at scale.

Due to the sensitive chemical and physical nature of oxalic acid, the phase equilibrium complexity of alcohols and esters involved, and the highly non-ideal behavior of the multi-component system with both homogeneous and heterogeneous azeotropes, the esterification step is a bottle-neck in the manufacturing of MEG from CO_2.

Reactive distillation (RD) is successfully applied to the esterification in this complex system to intensify process equipment as well as to move the chemical reactions beyond equilibrium limit due to the significant advantages of RD in conversion, selectivity and energy and capital savings with inherent advantageous integration of chemical, physical and thermal system challenges in one unit. Several alcohols have been investigated to identify the best candidate for optimized product and water removal mechanisms. The column is packed with solid acid catalyst using structured commercial packing in the reaction zone. The process is carried out with the select alcohol and optimized reactor and column configuration, system pressure, reflux and boilup ratios, alcohol to acid ratio, catalyst loading, column sizing and rating as well as relative dimension of reaction vs stripping and rectifying, temperature and species concentration profiles, material feeding positions, recycles and other operating conditions so to move the reactions to close to complete conversion with minimum side-product formation and higher than 98% di-ester yield.

A process has been developed for oxalic acid esterification and ester separation with minimized CAPEX and OPEX to ensure downstream hydrogenation step to produce fiber grade MEG with competitive economic advantages. The process is at pilot testing stage.

SOLAR THERMOCHEMICAL CONVERSION OF CO_2 INTO SYNTHETIC FUELS VIA FERRITE BASED REDOX REACTIONS

Rahul Bhosale[1,], Ivo Alxneit[2], Anand Kumar[1], LJP van den Broeke[1], Jamila Folady[1], Dareen Dardor[1], Shahd Gharbia[1], Mehak Jilani[1], Mashail Shaif Alfakih[1], Noura Nabil Elshrbajy[1], Mahsa Tarsad[1], Fatima Abdulaqder Almkhyari[1]*

[1]Department of Chemical Engineering, College of Engineering, Qatar University, PO Box 2713, Doha, Qatar.
[2]Solar Technology Laboratory, Paul Scherrer Institute, CH-5232 Villigen PSI, Switzerland.
**Corresponding Author: rahul.bhosale@qu.edu.qa*

Abstract

In this contribution, we report the synthesis $NiFe_2O_4$ and $CoFe_2O_4$ redox materials via sol-gel method. For the synthesis of these materials via sol-gel approach, the Ni, Co, and Fe precursors were first dissolved in ethanol with the help of sonic energy. Once the metal precursors were dissolved, propylene oxide (PO) was added dropwise to the well mixed solution as a gelation agent to achieve gel formation. As-prepared gels were aged, dried and subsequently calcined in presence air. The calcined powder obtained was characterized towards its phase/chemical composition, particle morphology, and specific surface area (SSA). Derived $NiFe_2O_4$ and $CoFe_2O_4$ redox materials were further investigated towards thermochemical splitting of CO_2 into solar fuels by performing several reduction/re-oxidation cycles using a thermogravimetric analyzer (TGA).

Introduction

Solar radiation is an essentially inexhaustible energy source that delivers about 100,000 TW to the earth. To harvest the solar radiation and to convert it effectively into renewable fuels from H_2O and captured CO_2 provides a promising path for a future sustainable energy economy [1]. Solar fuel production via metal oxide (MO) based thermochemical H_2O and/or CO_2 splitting reactions are considered as one of the promising new technologies for fulfillment of future energy requirement and steadying the greenhouse gas concentration in the earth's atmosphere. In comparison to the high temperature direct thermolysis of H_2O and/or CO_2 (needs more than 2500°C), the MO based thermochemical cycles are advantageous as: a) these cycles requires lower temperatures as compared to thermolysis, b) no explosive mixture formation as the production of H_2/CO and O_2 can be carried out in two different steps, and c) environmentally and thermodynamically more feasible in comparison with thermolysis [3-4].

Production of solar fuels via MO based thermochemical reactions is a two-step process. In the first step, the MO is reduced into a lower valence MO or metal with the help of solar energy. The reduced MO is further re-oxidized in the second step via H_2O and/or CO_2 splitting reactions. Among the many MOs investigated so far for solar fuel production, in recent years, research has been focused towards mixed ferrite materials [1-11]. These redox materials are appealing as their reduced form is a solid and hence the separation of a gaseous metal (or MO) and O_2, a challenging and compulsory step in cycles based on ZnO/Zn or SnO_2/SnO/Sn

[12-16] can be eliminated. A typical ferrite based solar thermochemical H_2O/CO_2 splitting process is presented in Fig. 1.

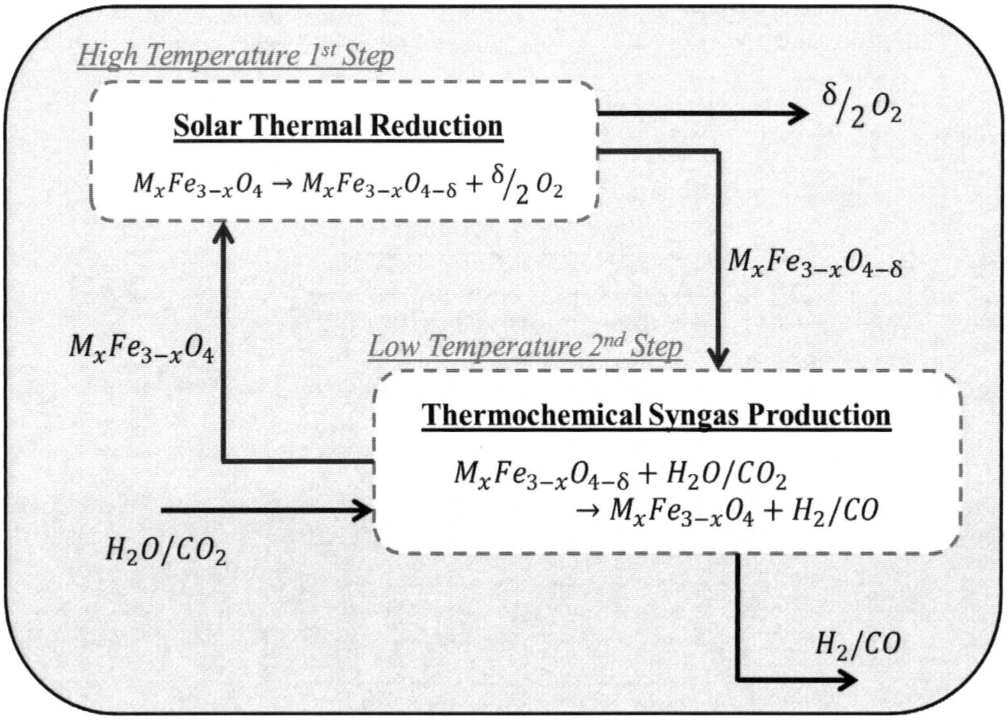

Fig.1. Typical ferrite based two-step solar thermochemical H_2O/CO_2 splitting process.

According to the results reported in previous investigations, it was observed that Ni-doped ($NiFe_2O_4$) and Co-doped ferrite materials ($CoFe_2O_4$) showed promising towards solar H_2 production via thermochemical water splitting process. Based on this background, this investigation reports the utilization of $NiFe_2O_4$ and $CoFe_2O_4$ redox materials towards solar thermochemical CO production via CO_2 splitting reaction. Respective ferrite materials were synthesized using sol-gel approach and the derived materials were characterized using various analytical methods such as powder x-ray diffractometer (XRD), BET surface area analyzer, scanning (SEM) and transmission electron microscopy (TEM), and an inductively coupled plasma spectrometer (ICP). Sol-gel derived $NiFe_2O_4$ and $CoFe_2O_4$ were examined towards their thermal reduction and CO_2 splitting ability by performing successive thermochemical cycles using a thermogravimetric analyzer (TGA).

Experimental

Synthesis of ferrite materials

For the synthesis of $NiFe_2O_4$ and $CoFe_2O_4$ redox materials via sol-gel method, the nitrate salts of Ni or Co, and Fe were added in ethanol and this mixture was sonicated to dissolve the metal salts in the solvent. Once the metal precursors were dissolved in ethanol, predetermined amount of propylene oxide (PO) was added dropwise as a gelation agent and the gel formation was achieved. As-synthesized gel was aged by keeping undisturbed for 24 h.

After aging, the gel was broken and dried using a vacuum oven operated at 100°C. The dried ferrite powder obtained was further calcined rapidly upto 600°C in air using a muffle furnace. Fig. 2 represents a typical synthesis route followed in this investigation for the production of $CoFe_2O_4$ and similar approach was used for the synthesis of $NiFe_2O_4$.

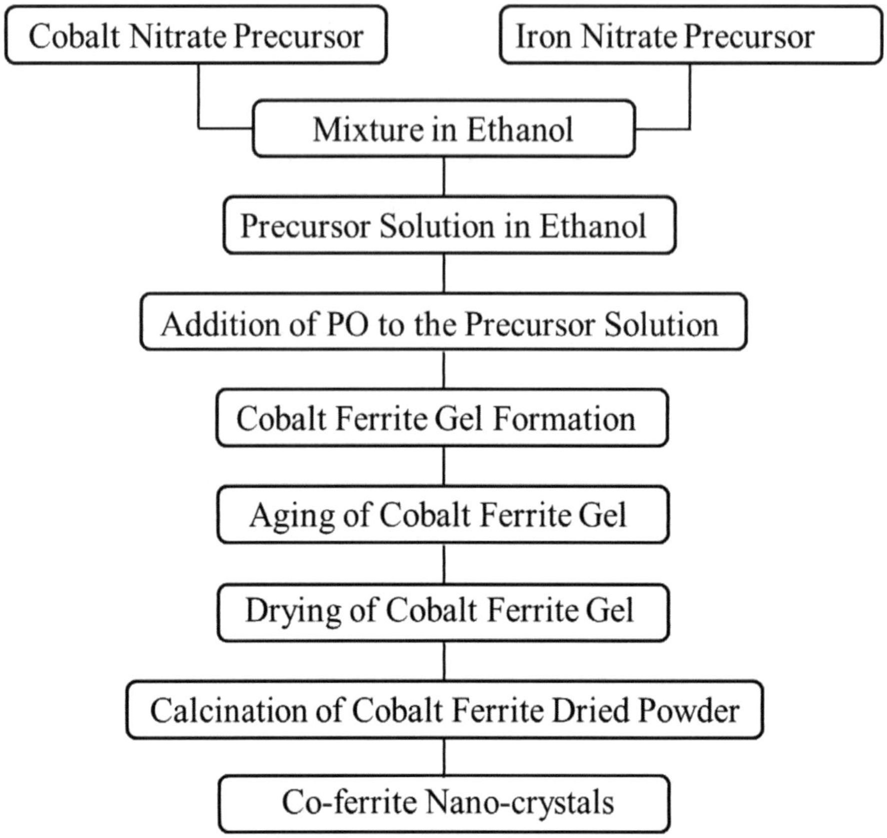

Fig.2. Sol-gel route used for the synthesis of ferrite materials.

Characterization of ferrite materials

Phase purity, crystallite size, morphology, specific surface area, and elemental composition of the sol-gel derived ferrite powders were analyzed using a Panalytical XPert MPD/DY636 powder X-ray diffractometer, a Zeiss Supra 55VP field-emission scanning electron microscope (SEM), a FEI – Tecnai G2 200kV transmission electron microscope (TEM), a BET surface area analyzer (Micromeritics, ASAP 2420), and an ICP (Thermo, iCAP 6500).

TGA set-up and procedure

The thermal reduction and the subsequent CO_2-splitting cycles were performed by using the $NiFe_2O_4$ and $CoFe_2O_4$ redox materials derived via sol-gel method using a thermogravimetric analyzer (TGA, Netzsch STA 409). Approximately 50 mg of ferrite powder was placed in an alumina crucible. Thermal reduction of ferrites was carried out at 1400°C for 60 minutes in presence of 100 ml/min of Ar as an inert gas. After completing the thermal reduction step, the temperature was lowered to 1000°C to perform the CO_2 splitting reaction. The CO_2 was admitted (50 ml of CO_2/min + 50 ml of Ar/min) for 30 min to induce the re-oxidation of the ferrites. The mass loss/increase observed is correlated with the O_2 release/uptake of the sample during the reduction/re-oxidation step. O_2 and CO evolution was determined by on-line gas chromatography (GC, VARIAN, CP-4900, Micro GC 2 channel system)) of the off gas flow.

Results and Discussion

Powder XRD analysis was performed to identify the phase composition of the sol-gel derived $NiFe_2O_4$ and $CoFe_2O_4$ redox materials. The XRD patterns for both materials calcined at 600°C in air are presented in Fig. 3. As per the results obtained, both ferrite materials showed strong XRD peaks/reflections and a well-defined spinel cubic structure with high degree of crystallinity. XRD reflections related to Impurities such as NiO, CoO, pure metals i.e. Ni, Co, or Fe were not observed. When compared with the ICDD patterns of pure iron oxide (Fe_3O_4), it was understood that due to the inclusion of Ni or Co metal into spinal crystal structure of iron oxide, the diffraction patterns of derived $NiFe_2O_4$ and $CoFe_2O_4$ redox materials were observed to be shifted as compared to the pure iron oxide. For further confirmation of compositional purity, ICP analysis was also performed and the results obtained indicate that phase pure $NiFe_2O_4$ and $CoFe_2O_4$ redox materials were produced via sol-gel approach utilized in this investigation (Table 1).

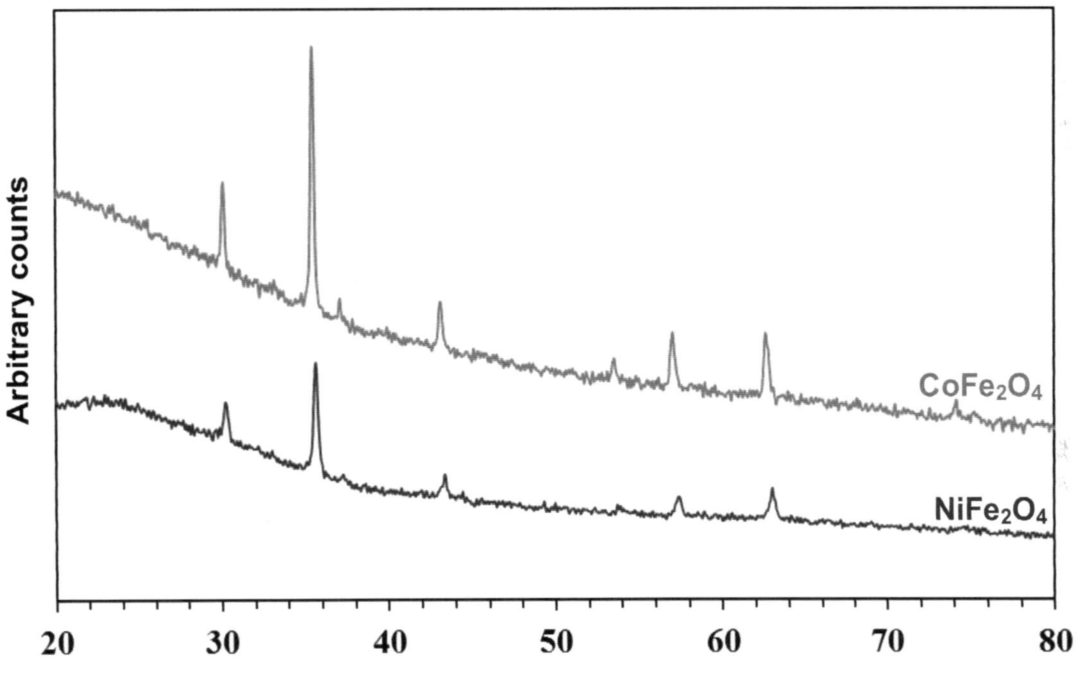

Fig. 3. Powder XRD patterns for $NiFe_2O_4$ and $CoFe_2O_4$ redox materials.

Crystallite size of single crystalline domains of sol-gel derived $NiFe_2O_4$ and $CoFe_2O_4$ redox materials were calculated using Scherrer equation. The crystallite size for $NiFe_2O_4$ was observed to be 29 nm whereas for $CoFe_2O_4$ it was equal to 31 nm. Further information about the particle size and morphology of the derived ferrite materials was gathered by performing SEM and TEM analysis experiments. The results obtained via SEM and TEM analysis indicate that the synthesis of ferrites via sol-gel method yields into nano-crystals of $NiFe_2O_4$ and $CoFe_2O_4$ redox materials. The specific surface area (SSA) and porosity of the ferrite materials was determined using a BET surface area analyzer. In case of $NiFe_2O_4$, the SSA was equal to 36.52 m^2/g and pore volume was observed to be 0.0312 cm^3/g. Likewise, in case of $CoFe_2O_4$, the SSA and pore volume were observed to be equal to 37.43 m^2/g and 0.0722 cm^3/g.

Table 1: ICP analysis results for sol-gel derived ferrite materials.

Material	Ni:Fe ratio (prepared)	Ni:Fe ratio (calcined)	Final Composition
Ni-ferrite	1:2	1:2	$NiFe_2O_4$
Co-ferrite	1:2	1:2	$CoFe_2O_4$

The redox relativity and thermal stability of the sol-gel derived $NiFe_2O_4$ and $CoFe_2O_4$ redox materials during thermochemical CO2 splitting cycles was examined by performing multiple thermal reduction and re-oxidation experiments using a thermogravimetric analyzer (TGA). Prior conducting actual thermochemical CO_2 splitting experiments using $NiFe_2O_4$ and $CoFe_2O_4$ redox materials, a baseline run was performed under identical experimental conditions in absence of the ferrite powder (using empty alumina crucible). This allows subtracting artifacts due to buoyancy effects or caused by changing the gas composition.

Four consecutive thermochemical cycles were performed to investigate the thermal reduction and CO_2 splitting ability of the sol-gel derived $NiFe_2O_4$ and $CoFe_2O_4$ redox materials. The results obtained during the TGA analysis are reported in Fig. 4. Thermal reduction step was carried out at 1400°C for 60 min in presence of pure inert Ar. On the other hand, CO_2 splitting experiment was performed at 1000°C for 30 min in presence of a gas mixture consists of CO_2 and Ar (50:50 molar ratio). According to the results reported in Fig. 4, it can be seen that the redox capacity of both sol-gel derived $NiFe_2O_4$ and $CoFe_2O_4$ redox materials does not deteriorate i.e. approximately the same mass loss and mass gain is observed in all thermochemical cycles.

Amounts of evolution of O_2 and CO during in consecutive thermochemical cycles performed using sol-gel derived $NiFe_2O_4$ and $CoFe_2O_4$ redox materials were determined with the help of % weight changes (TGA analysis) and the data obtained via gas chromatography. During the 1st thermal reduction step, both ferrites produced a very high amount of O_2 (1.37% mass loss in case of $NiFe_2O_4$ and 1.37% mass loss in case of $CoFe_2O_4$). This disproportionally large weight loss occurs due to the desorption of physisorbed water from the ferrite samples. In contrast to this, for the 2nd, 3rd, and 4th thermochemical CO_2 splitting steps, the % mass loss and mass gain by both ferrite materials were observed to be constant. In case of $NiFe_2O_4$, an average of 269 μmol/g·cycle O_2 and 360 μmol/g·cycle of CO were produced with CO to O_2 ratio equal to 1.34. Likewise, sol-gel derived $CoFe_2O_4$ produce 273 μmol/g·cycle O_2 and 424 μmol/g·cycle of CO with CO to O_2 ratio equal to 1.55. Obtained results indicate that the thermal reduction ability of both ferrites is identical; however, $CoFe_2O_4$ is capable of producing higher amount of CO as compared to $NiFe_2O_4$. Furthermore, these results also shows that sol-gel derived $NiFe_2O_4$ and $CoFe_2O_4$ redox materials are capable of producing higher amounts of solar fuels as compared to that of the previously investigated redox materials such as doped ceria redox materials [12-21].

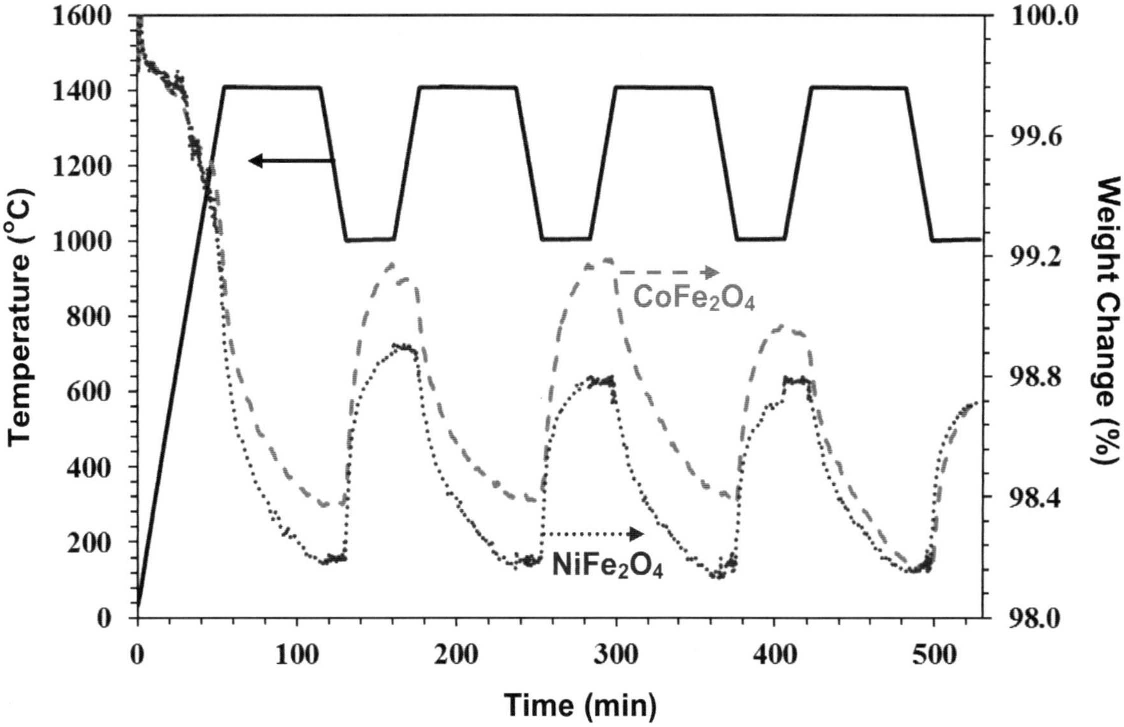

Fig. 5. TGA of CO_2-splitting experiments performed using sol-gel derived $NiFe_2O_4$ and $CoFe_2O_4$ redox materials.

Conclusions

In this study, $NiFe_2O_4$ and $CoFe_2O_4$ redox materials were effectively derived via sol-gel method by using nitrate salts of Ni, Co, and Fe, ethanol as solvent, and PO as gelation agent. Characterization performed with the help of powder XRD indicate that the sol-gel derived $NiFe_2O_4$ and $CoFe_2O_4$ redox materials are phase pure in composition with absence of any oxide or metal impurities. The crystallite/particle size and material morphology analyzed via quantitative XRD and SEM/TEM reveal formation of $NiFe_2O_4$ and $CoFe_2O_4$ nanocrystals. The solar fuel production capacity of both ferrite materials were further examined in the temperature range of 1000_oC to 1400^oC using a thermogravimetric analyzer. TGA and GC results showed that thermal reduction capacity of both ferrites is identical, however CO2 splitting ability of $CoFe_2O_4$ was observed to be reasonably higher as compared to the $NiFe_2O_4$ redox material. Nevertheless, both materials produced highest amounts of O_2 and CO as compared to previously investigated redox materials (at similar operating conditions).

Acknowledgments

The authors gratefully acknowledge the financial support provided by the Qatar University (QUUG-CENG-CHE-13/14-4), Indo-Swiss Joint Research Program (ISJRP, grant #138852), and Swiss Federal Office of Energy.

References

1. Bhosale, R., Alxneit, I, van den Broeke, Leo L. P., Kumar, A., Jilani, M., Gharbia, S., Folady, J., and Dardor, D. (2014), "Sol-gel synthesis of nanocrystalline Ni-Ferrite and Co-Ferrite redox materials for thermochemical production of solar fuels", *Proceedings of Materials Research Society Symposium* (Accepted – In Press).
2. Bhosale, R., Shende, R., and Puszynski, J. (2010), "H_2 generation from thermochemical water-splitting using sol-gel synthesized ferrites ($M_xFe_yO_z$, M= Ni, Zn, Sn, Co, Mn, Ce)", *Proceedings of AIChE Annual Meeting*.
3. Shende, R., Puszynski, J., Opoku, M., Bhosale, R. (2009), "Synthesis of novel ferrite foam material for water-splitting application", *Proceedings of NSTI-Nanotech Conference & Expo. ISBN 978-1-4398-1782-7,* **201**, 1.
4. Kodama, T., Gokon, N., and Yamamoto, R. (2010), "Thermochemical two-step water splitting by ZrO_2-supported $Ni_xFe_{3-x}O_4$ for solar hydrogen production", Solar Energy, **82**, 73.
5. Scheffe, J., Li, J., and Weimer, A. (2010), "A spinel ferrite/hercynite water-splitting redox cycle", *International Journal of Hydrogen Energy*, **35**, 3333.
6. Bhosale, R., Shende, R., and Puszynski, J. (2010), "H_2 generation from thermochemical water splitting using sol-gel derived Ni-ferrite", *Journal of Energy and Power Engineering*, **4**, 27.
7. Bhosale, R., Shende, R., and Puszynski, J. (2010), "H_2 generation from thermochemical water-splitting using sol-gel synthesized Zn/Sn/Mn-doped Ni-ferrite", *International Review of Chemical Engineering*, **2**, 852.
8. Bhosale, R., Shende, R., and Puszynski, J. (2012), "Thermochemical water-splitting for H_2 generation using sol-gel derived Mn-ferrite in a packed bed reactor", *International Journal of Hydrogen Energy*, **37**, 8223.
9. Bhosale, R., Khadka, R., Shende, R., and Puszynski, J. (2011), "H_2 generation from two-step thermochemical water-splitting reaction using sol-gel derived $Sn_xFe_yO_z$", *Journal of Renewable and Sustainable Energy*, **3**, 063104-1.
10. Bhosale, R., Shende, R., and Puszynski, J. (2012), "Sol-gel derived $NiFe_2O_4$ modified with ZrO_2 for hydrogen generation from solar thermochemical water-splitting reaction", *Proceedings of Materials Research Society Symposium*, 1387.
11. Fresno, F., Yoshida, T., Gokon, N., Fernandez-Saavedra, R., and Kodama, T. (2010), "Comparative study of the activity of nickel ferrites for solar hydrogen production by two-step thermochemical cycles", *International Journal of Hydrogen Energy*, **35**, 8503.
12. Loutzenhiser, P., and Steinfeld, A. (2011), "Solar syngas production from CO_2 and H_2O in a two-step thermochemical cycle via Zn/ZnO redox reactions: Thermodynamic cycle analysis", *International Journal of Hydrogen Energy*, **36**, 12141.
13. Stamatiou, A., Loutzenhiser, P., and Steinfeld, A. (2010), "Solar syngas production from H_2O and CO_2 via two-step thermochemical cycles based on Zn/ZnO and FeO/Fe_3O4 redox reactions: Kinetic analysis", *Energy & Fuels*, **24**, 2716.
14. Loutzenhiser, P., Meier, A., and Steinfeld, A. (2010), "Review of the two-step H_2O/CO_2-splitting solar thermochemical cycle based on Zn/ZnO redox reactions", *Materials*, **3**, 4922.

15. Charvin, P., Abanades, F., Lemont, F., and Flamant, G. (2008), "Experimental study of SnO_2/SnO/Sn thermochemical systems for solar production of hydrogen", *AIChE Journal*, **54**, 2759.
16. Abanades, F. (2012), "CO_2 and H_2O reduction by solar thermochemical looping using SnO_2/SnO redox reactions: Thermogravimetric analysis", *International Journal of Hydrogen Energy*, **37**, 8223.
17. Scheffe, J., Jacot, R., Patzke, G., and Steinfeld, A. (2013), "Synthesis, characterization and thermochemical redox performance of Hf, Zr and Sc doped ceria for splitting CO_2", *Journal of Physical Chemistry C*, **177**, 24104.
18. Chueh, W. C., Falter, C., Abbott, M., Scipio, D., Furler, P., Haile, S., and Steinfeld, A. (2010), "High-flux solar-driven thermochemical dissociation of CO_2 and H_2O using nonstoichiometric ceria", *Science*, **330**, 1797.
19. Le Gal, A., and Abanades, S. (2011), "Catalytic investigation of ceria-zirconia solid solutions for solar hydrogen production", *International Journal of Hydrogen Energy*, **36**, 4739.
20. Abanades, S., and Gal, A. L. (2012), "CO_2 splitting by thermo-chemical looping based on $Zr_xCe_{1-x}O_2$ oxygen carriers for synthetic fuel generation", *Fuel*, **102**, 180.
21. Le Gal, A., Abanades, S., and Flamant, G. (2011), "CO_2 and H_2O splitting for thermochemical production of solar fuels using nonstoichiometric ceria and ceria/zirconia solid solutions", *Energy & Fuels*, **25**, 4836.

pH Effect on growth mechanism of *Nannochloropsis Oculata* in the study of CO_2 capture process:

M. Ilyas*, S. A. Razzak
Chemical Engineering Department, King Fahd University of Petroleum and Minerals KSA

ABSTRACT

As the greenhouse gases are the major environmental concern worldwide and scientist consider these gases responsible for global warming. CO_2 is a major constituent, produced by burning fossil fuel worldwide, lot of research is being done for the CO_2 mitigation purpose to preserve planet environment. Trends are shifting towards renewable energy sources in order to minimize CO_2 emission. Biological CO_2 capture methods using photo-bioreactors by growing microalgae, which is potential technology being explored worldwide due to its advantage of CO_2 capture and biofuel production as well. As in the Middle East countries especially in Saudi Arabia region has large power plants as CO_2 source and wastewater as nutrients source for microalgae growth and higher solar irradiations as extra benefit for photosynthesis. In this work with the purpose of CO_2 reduction, *Nannochloropsis Oculata* was cultured in batch photo-bioreactor with CO_2 concentrations of 2%, 4%, 6%, 8%, 10% and 12% mixed with air to study the pH effect on growth kinetics, biomass productivity and CO_2 reduction with the change in CO_2/Air feed composition. Results shows, pH is the crucial factor for *Nannochloropsis Oculata* culture process which is directly related to CO_2 feed composition and growth media. The results of this study will be helpful to define optimum operating parameters for culture developing of *Nannochloropsis Oculata* with maximum CO_2 utilization and biomass production which will contribute towards reduction of environmental concerns caused by CO_2.

Acknowledgement: The authors would like to acknowledge the support provided by King Abdul-Aziz City for Science and Technology (KACST) Saudi Arabia for funding this research work through project funding process.

**Author to whom correspondence should be addressed.*

Engr. Muhammad Ilyas
E-mail: ilyasaali@kfupm.edu.sa; Tel: +966-3-860-3247; Mobile: +966-55-784-5583

1. Introduction:

As due to the increase of global warming, which can be mainly due to the elevated CO_2 level in the atmosphere, the United Nations signed the Kyoto Protocol (1997) with the objective to reduce greenhouse gases by 5.2% on the basis of the emission in year 1990. Main CO_2 mitigation techniques investigated are chemical reaction-based approaches and biological CO_2 mitigation method. The chemical reaction-based CO_2 fixation typically consists of three procedures of separation, transportation, and sequestration of CO_2 (Gupta and Fan 2002). As this method is relatively costly and more energy-consuming (Stewart and Hessami, 2005), so it is therefore necessary to develop cost-effective and suitable alternatives method (Herzog H, 1997).

Worldwide energy demand is ever increasing while conventional fossil fuel reserves are depleting, also fossil fuel causes environmental problems. We need to produce renewable energy sources in order to reduce CO_2 emissions produced by fossil fuels. Finding of efficient methods of CO_2 capture which are environmental friendly. Microalgae have potential to fix CO_2, wastewater treatment and biofuel production as well (Amit Kumar et al., 2010).

Biological CO_2 mitigation is more attractive as an alternative because it leads to production of biomass energy as byproduct in the process of CO_2 fixation through photosynthesis (Packer, 2009). Mainly CO_2 is converted into algal biomass by the process of photosynthesis in the presence of light source in photo bioreactor (Zeiler et al, 1995). After processing the treatment(biological, physical or thermochemical) the biomass generated can be used for production of green fuel like biodiesel, methane or other fuels, food and other co-products (Stephens et al., 2010;Brennan and Owende 2009). Also using wastewater as a growth medium provides multiple advantages, such as the removal of nitrates, phosphates and even some heavy metals from the wastewater whilst providing nutrients for the algae (Mata et al., 2010;de-Bashan & Bashan, 2010)

2. Problem and Objective of study:

pH based study of microalgae specie (*Nannochloropsis Oculata*) for the purpose of CO_2 capture and biomass production by applying different operating conditions using photobioreactor. For this porpose microalgae speice growth was observed under different conditions of CO_2/Air mixture by providing constant flouriscent light on the either side of photobioreactor. Light, pH, temprature was monitored on daily basis. Under these operating conditions growth was studied with the following objectives.

- To investigate the growth mechanism of *Nannochloropsis Oculata*
- To investigate pH effect on Air/CO_2 ratios in the growth of *Nannochloropsis Oculata*
- Finding the CO_2 bio fixation rate for different Air/CO_2 input conditions

3. Experimental Methodology:

3.1 Batch photo-bioreactors setup:

Photo-bioreactor for the experiments on CO_2 mitigation for batch cultures is shown in figure. The airstreams with different CO_2 concentrations was mixed using gas mixer and inlet, outlet concentrations determined by CO_2 sensor. External illumination of light is provided on either side.

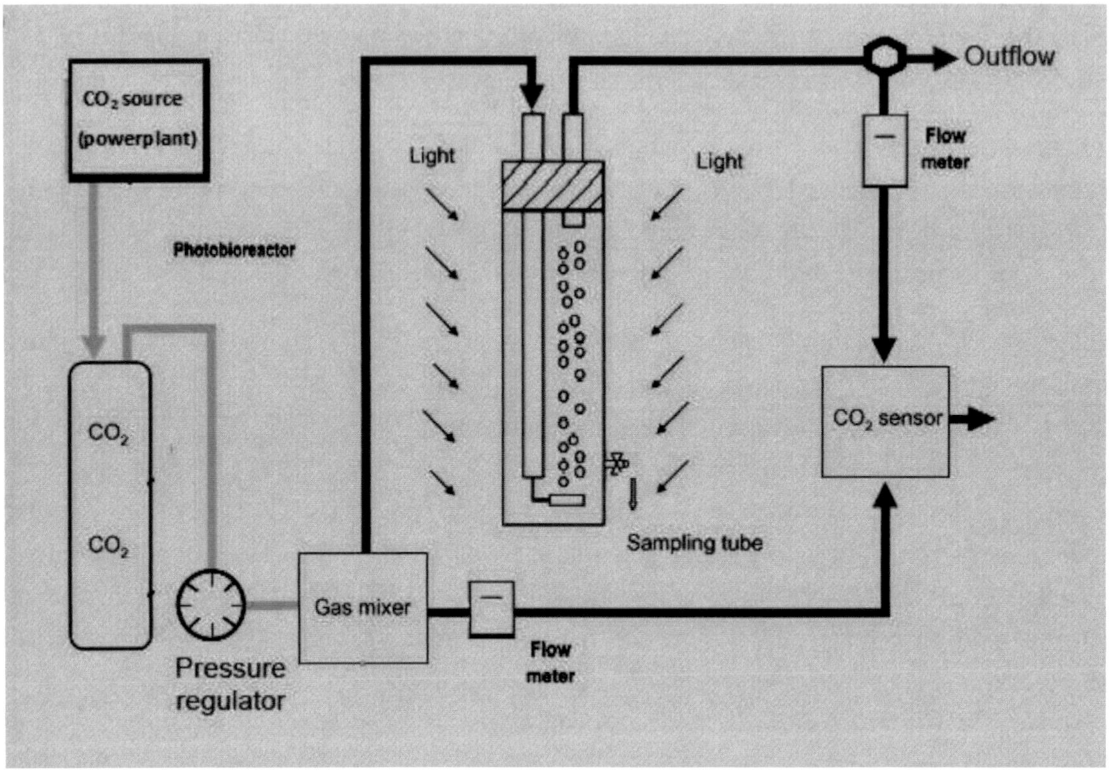

Figure 1: Schematic diagram of experimental setup

3.2 Growth kinetics study:

Direct measurement by microscope (Fisher Scientific™ Micromaster Microscope) was employed to measure the cell concentration. Dry biomass was calculated by filtration method by applying vacuum. Specific growth rate and biomass productivity calculations on per day basis were made using the following relation.

$$\mu = \frac{ln(X_m/X_0)}{t_n - t_0} \quad \& \quad p = \frac{X_m - X_0}{t_n - t_0}$$

Where X_m and X_0 are biomass calculated and t_n and t_0 are the time of sample measurement. The pH value of cultures was monitored on daily basis using desktop pH meter (Fisher Scientific Accumet® Basic AB15 Plus pH Meter). The CO_2 fixation rate was calculated from the analysis of the biomass and growth kinetics data as shown in equation,

$$R_c = C_c\, X \left(\frac{X_m - X_0}{t_n - t_0}\right) X \left(\frac{M_{CO_2}}{M_C}\right)$$

Where C_c carbon content of microalgae species determined using TOC analyzer (Teledyne Tekmar® Torch Combustion TOC/TN Analyzer): M_{CO_2} is the molecular weight of CO_2: M_c is the molecular weight of carbon: where X_m *and* X_0 are the biomass at time t_n and t_0 respectively.

4. Results and Discussions:

4.1 Growth mechanism of *Nannochloropsis Oculata*:

The growth of *Nannochloropsis Oculata* is divided into three regions as shown in the Figure 2. First region is initial growth region called "lag region" specified as "A" in figure. When the culture is incubated in fresh media then it takes some time for the cells to adjust with new environment. The second region of growth is exponential growth or logarithmic growth specified as "B" in the figure. During this stage they utilize maximum of their nutrients and food elements which mean more CO_2 utilization to produce daughter cells.

The third region which is known as stationary region is described as part "C" of the curve in figure. There are several factors which are considered responsible for stationary growth i) exhaustion of nutrients ii) while at low cell densities CO_2 lowers the pH and growth is depressed iii) light limitation due to self-shading effect of cells iv) accumulation of waste material iv) toxic metabolic compounds.

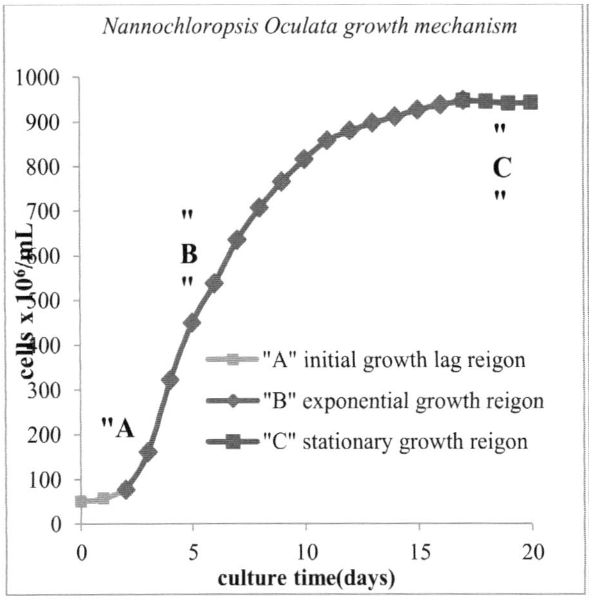

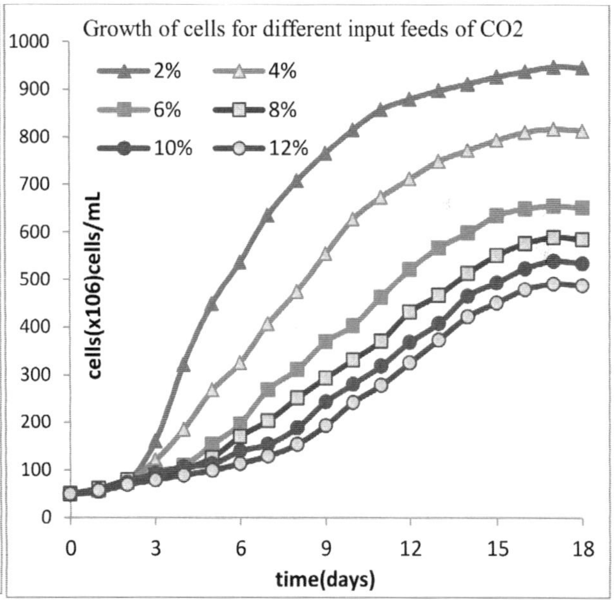

Figure 2: Growth rate mechanism in general. Figure 3: Growth kinetics in cells for diff.CO_2

4.2 Effect of CO_2 concentration (pH) on growth:

Figure 3 shows the effect of CO_2 concentration towards the cell growth and biomass yield of *Nannochloropsis Oculata*. From the figure *Nannochloropsis Oculata* was found to grow in lower CO_2 concentrations favorably where growth rate is accelerated by shifting towards lower CO_2 concentration due to pH effect. As an example the growth rate of *Nannochloropsis Oculata* increased from 2.05 day^{-1} to 2.54 day^{-1} (increment of 23.9%) when CO_2 concentration saturated with air was shifted from 12% to 2% CO_2. Similar results were found for multiplication of cell concentration. In addition to growth rate acceleration, the biomass productivity is also increased from 0.034 $mgL^{-1}day^{-1}$ to 0.094 $mgL^{-1}day^{-1}$ with an increment of 180% when CO_2 concentration shifted from 2% to 12% CO_2. From these observations we can conclude that specified microalgae species cannot tolerate higher CO_2 concentration due to change in pH which becomes acidic with higher CO_2 concentration which is not favorable for growth.

4.3 CO_2 bio-fixation rate:

The CO_2 biofixation rate was determined using the equation described in methodoligy and results for six differents CO_2 concentrations are shown in figure 5. The maximum CO_2 biofixation rate found was 0.086 g L^{-1} d^{-1} with CO_2 concentration of 2% at day 5 of culture (figure 4). Simillar trend of CO_2 biofixation was observed for all CO_2 concentrations. CO_2 fixation rate is first incresed as the growth procedes with msximum growth rate it reaches to maximum values afer that with the decline in growth CO_2 fixation is also decresed. CO_2 biofixation abilities under different CO_2 concentrations from 2% to 12% CO_2 concentration and performed the best CO_2 fixation at 4% CO_2.

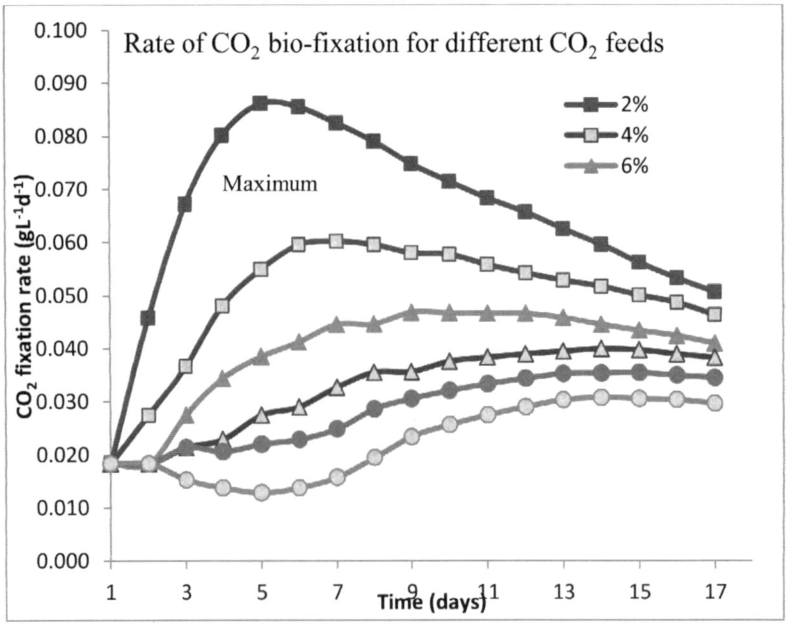

Figure 4: CO_2 biofixation rate for different CO_2 concentrations.

4.5 Optimum CO_2/Air feed for maximum growth:

Table represent the maximum biomass, maximum productivity and CO_2 fixation rate with percent increment of these values with shift of CO_2 concentration from 2% - 12%. From these observations of results it shows decline with increase CO_2 from 2% to 12% mixed with air. Overall productivity, biomass and CO_2 fixation rate observed is maximum up to 2% CO_2 exceeding this limit results in decline of growth.

Table 1: Maximum biomass, productivity and CO_2 fixation rate with percent increment for Nannochloropsis Oculata at different CO_2 concentrations.

CO_2 conc.	**Max. Biomass**	**Biomass Increment**	**Max. Productivity**	**Yield Increment**	**Specific Growth**	**R_{CO2} fixation**
%	X_{max} g L^{-1}	% from 12%-2%	P_{max} g L^{-1} d^{-1}	% bottom to top	R_{max} d^{-1}	R_{maxCO2} g L^{-1} d^{-1}
2	1.02	61.90	0.094	176.47	2.544	0.086
4	0.94	49.21	0.066	94.12	2.46	0.06
6	0.84	33.33	0.051	50.00	2.35	0.047
8	0.79	25.40	0.044	29.41	2.28	0.041
10	0.72	14.29	0.039	14.71	2.19	0.036
12	0.63	0.00	0.034	0.00	2.06	0.031

4.6 Effect ofpH on CO_2/Air Optimization :

The pH of the medium without microalgae culture is around 7 when culture was incubated with microalgae sample and CO_2 with different concentration was fed to the culture its pH drops due to CO_2 absorption in aqueous phase, with time it again increases as the growth is started to increase. Overall average of pH for 18-days operation is in decreasing order with increasing order of CO_2 % mixed with air but this decrease is not significant which showed that aqueous CO_2 dissolved in medium is almost constant it could not dissolve any further increase in CO_2 inlet stream beyond 2% of CO_2 as indicated from the results. This showed that most of the CO_2 at influent directly flowed out of the culture medium reactor when CO_2 concentration was increased beyond 2% of CO_2 in air mixed stream.

The efficiency of CO_2 removal in closed photobioreactor system is mainly dependent on the following i: microalgae species ii: CO_2 concentration iii: medium selection iv: photobioreactor selection v: residence time of gas streams in culture (Cheng et al., 2006). The dependence of CO_2 fixation on microalgae species may be due to the potential of growth of that specie and CO_2 metabolism activity.

5 Conclusion:

From these results it is concluded that CO_2 can be reduced by growing different microalgae species with different CO_2 compositions which reveals that biological CO_2 fixation in the countries like Saudi Arabia which has vast CO_2 production sources like power plants, wastewater treatment plants as source of nutrients for microalgae and round the year sun light and coastal area of different water compositions is more viable and economical towards the green and CO_2 free environment. From these experimental results for experimental specie of microalgae it is clear that it has CO_2 fixation potential and increasing growth rates within specified optimal range of temperature, light intensity and in controlled pH conditions. In this case of *Nannochloropsis Oculata* its shows more growth with lower CO_2 mixture with air as compared to higher CO_2 values. For investigations of more efficient CO_2 reduction species specifically used for these experiments it's clearly identified that *Nannochloropsis Oculata* has CO_2 reduction potential with biomass production which leads to valuable products. From these experiments we conclude that biological CO_2 reduction is a potential field of technology with its additional advantages of green fuel production from dry biomass and waste water treatment as well.

6 References:

1 Gupta H, Fan LS (2002) Carbonation–calcination cycle using high reactivity calcium oxide for carbon dioxide separation from flue gas Ind Eng Chem Res 41:4035–4042.

2 Stewart, C. & Hessami, M.-A 2005.A study of methods of carbon dioxide capture and sequestration - the sustainability of a photosynthetic bioreactor approach. Energy Conversion and Management,46(2005), pp.403-20

3 Herzog, H.; Drake, E.; Adams, E. CO_2 Capture, Reuse, and Storage Technologies for Mitigating Global Climate Change. A White Paper; Final Report No DE-AF22-96PC01257; U.S. Department of Energy: Washington, DC, Jan 1997

4 Packer, M., 2009. Algal capture of carbon dioxide, biomass generation as a tool for greenhouse gas mitigation with reference to New Zealand energy strategy and policy, Energy Policy, 37(9), pp.3428-37

5 Chang, J.S., Chen, C.Y., Yeh, K.L., Aisyah, R., Lee, D.J., 2011. Cultivation, photobioreactor design and harvesting of microalgae for biodiesel production: a critical review. Bioresource Technology 102, 71–81.

6 Amit Kumar1,2,3, Sarina Ergas4, Xin Yuan1, Ashish Sahu5, Qiong Zhang, Enhanced CO_2 fixation and biofuel production via microalgae: recent developments and future directions.Trends in biotechnology Volume 28, Issue 7, July 2010, Pages 371–380

7 Stephens, E. et al., 2010. Future prospects of microalgal biofuel production systems. Trends in Plant Science, xx(xx), pp.xx-xx.

8 Brennan, L. & Owende, P., 2010. Biofuels from microalgae, A review of technologies for production, processing, and extractions of biofuels and co-products. 14(2010), pp.557-77

9 Mata, T.M., Martins, A.A. & Caetano, N.S., 2010. Microalgae for biodiesel production and other applications: A review. Renewable and Sustainable Energy Reviews, 14(2010), pp.217-32

10 de-Bashan, L.E. & Bashan, Y., 2010. Immobilized microalgae for removing pollutants: Review of practical aspects, Bioresource Technology, 101(2010), p.1611–1627

Feasibility Study on CO_2 Storage and Waste Valorization for Cement Replacement using Steelmaking Fine Wastes

Kuan-Wei Chen , Pen-Chi Chiang

[a] Graduate Institute of Environmental Engineering, National Taiwan University, Taiwan (R.O.C.)

Abstract

Although the steel and cement industries are ubiquitous and beneficial to the world economy, great amounts of CO2 and industrial wastes like steelmaking slags are generated during their manufacturing processes. Recently, several research groups have actively worked on exploring cost-benefit alternative for CO_2 mitigation and waste valorization. The process of CO_2 fixation from large point sources has performed in rotating packed bed due to its higher mass-transfer rate. In this regard, the objective of this study was to investigate the potential use of carbonated product as partial cement replacement material. Fine wastes from steelmaking process were employed as feeding materials in this study.

First, accelerate carbonation of BOF slag in rotating packed bed under ambient temperature and pressure. Thermo gravimetric – mass spectrometry (TG-MS), Fourier transform infrared (FT-IR) spectroscopy, X-ray diffraction (XRD) and scanning electronic microscope with energy dispersion spectrometry (SEM-EDS) were used to characterize wastes before and after carbonation.

In addition, the hydration properties of cement paste with carbonated wastes including the standard consistency, setting time, autoclave expansion and heat of hydration were studied. Moreover, flow test, strength development, drying shrinkage and sulfate attack were observed on mortar specimens containing additional sand.

The results indicated that the effect of reaction temperature (e.g., 30 and 60 ^{o}C) on carbonation conversion of wastes was not significant. In addition, the compressive strength for blended cement with the 10% replacement of carbonated waste was found to be superior to that with the fresh waste.

It was thus concluded that the carbonated wastes may be feasible to partial replacement of the cement.

Keywords: steelmaking wastes, waste valorization, mineral carbonation, compressive strength, blended cement

A Novel Fluidized Bed Hydrator for Steam Reactivation of Calcium-based CO_2 Sorbents: Parametric Studies on Operating Velocity and Effect of Calcination Conditions

Alan Wang, Niranjani Deshpande, William Wang and L.-S. Fan*
William G. Lowrie Department of Chemical and Biomolecular Engineering
The Ohio State University
Columbus, Ohio 43210
U.S.A.
*To whom correspondence should be addressed.

Naturally occurring calcium sorbents, like lime and limestone, have been used for flue gas desulfurization for a number of years with success. Furthermore, calcium oxide (CaO)-based chemical looping systems have been proposed to control large point sources of carbon dioxide (CO_2) emissions, specifically coal-fired power plants. The deactivation of CaO sorbent for cyclic carbonation-calcination absorption processes has resulted in CaO hydration being considered as a promising approach to regenerate the sorbent. The three step Carbonation Calcination Reaction (CCR) process developed at the Ohio State University (OSU) proposes to use high temperature steam hydration as an intermediate step in the calcium loop. The concept has been demonstrated at various lab and bench scale operations. A high temperature steam hydration reactor was designed and preliminary results as proof of concept experiments have been published.

In the present study, further parametric test results are presented on the novel fluidized bed reactor. The inclusion of mechanical agitation is necessary to overcome the inherent operational challenges associated with the fluidization of the extremely cohesive fine calcium hydroxide ($Ca(OH)_2$), a known Geldart Type C particle. The effect of the operating gas velocity (U_g) in the reactor on both the solid and steam conversion has been investigated. The reaction rate is higher at higher U_g, and the highest conversion of 83% was obtained with 30 min residence time. However, higher U_g results in loss in the steam utilization or increase in waste steam. Thus, the optimization of operating velocity, with respect to high reaction rate and low steam requirement, will be discussed. Furthermore, a parametric study has been conducted for the effect of calcination conditions (specifically calcination temperature, time, and reactor type) on the sorbent physical morphology and the subsequent hydration reaction. Sorbent reactivity and morphology was tested using thermogravimetric (TGA) and Brunauer–Emmett–Teller (BET) analysis respectively. Calcination temperature is found to be the single most important factor for sorbent reactivity towards hydration. In addition, the calciner design greatly affects the sorbent reactivity. Lastly, the data suggests that the hydration mechanism depends strongly on the surface kinetics, a trend that is similar to the carbonation reaction, in other words, sorbents with larger surface area and pore volume exhibits more favorable reaction rates for both the carbonation and hydration reactions. From a process perspective, in the three-step CCR process, if both the carbonation and hydration mechanisms exhibit a strong dependence on the calcination conditions due to the sintering effects, an emphasis needs be placed on the optimal calciner design operating conditions to produce the most reactive sorbent.

Rational Design of Nano-Structured Oxygen Carriers for Chemical Looping Conversion of Methane

Fanxing Li
Department of Chemical and Bimolecular Engineering
North Carolina State University
Raleigh, NC 27695
* Email: fli5@ncsu.edu

As the primary component of natural gas, methane is among the most abundant organic compounds on earth. The current approach for liquid fuel and chemical production from methane involves methane reforming in the presence of gaseous oxidants such as steam, oxygen, and/or carbon dioxide. Although reforming based methods, particularly steam methane-reforming, have been successfully utilized in commercial processes, the efficiencies of state-of-the-art reforming processes are limited due to the high steam to methane ratio required by the reforming catalysts and/or the needs for energy intensive air separation operations. The current study investigates oxygen carriers, a.k.a redox catalysts, for converting methane into syngas and/or carbon dioxide in a chemical looping scheme. A number of redox catalysts composed of a primary metal oxide and mixed ionic-electronic conductor (MIEC) support, assembled as sub-micron composite or nano-sized core-shell particles, are synthesized and characterized. The mechanisms of the corresponding redox reactions are also investigated. Experimental results indicate that the core-shell redox catalysts are significantly more active than conventional oxygen carriers. Strategies for rational design and optimization of redox catalysts for chemical looping combustion and reforming schemes are also discussed based on these findings.

Encapsulated iron oxide nanoparticles for chemical looping based hydrogen production

Davood Hosseini, Christoph Müller
Laboratory of Energy Science and Engineering,
Department of Mechanical and Process Engineering, ETH Zurich, Switzerland

Chemical looping (CL) based hydrogen production involves a two-step process in which in the first step a synthesis gas reduces a metal oxide (Me_xO_y) oxygen carrier (Eqs. 1,2). Subsequently, the reduced metal oxide is re-oxidized using steam to produce a pure stream of H_2 (Eq. 3).

$$H_2 + Me_xO_y \leftrightarrow H_2O + Me_xO_{y-1} \quad (1)$$

$$CO + Me_xO_y \leftrightarrow CO_2 + Me_xO_{y-1} \quad (2)$$

$$H_2O + Me_xO_{y-1} \leftrightarrow H_2 + Me_xO_y \quad (3)$$

An appropriate oxygen carrier for the CL-based H_2 production must possess the following characteristics[1]:

(i) Thermodynamics that favor the complete conversion of synthesis gas to CO_2 and H_2O.

(ii) Regenerable with steam with a high 'hydrogen to steam' ratio.

(iii) Fast and stable kinetics over multiple cycles.

In this regard iron oxide is arguably the most promising material due to favorable thermodynamics, wide availability and being environmentally benign[1,2]. However, the lack of sufficient redox stability of pure Fe_2O_3 at high temperatures (700-1000 °C) is a major concern. Bohn et al.[3] demonstrated that unsupported Fe_2O_3 deactivates after only a few redox cycle. A possible solution to stabilize the redox performance of iron oxide is by supporting it on a high Tammann temperature material, such as SiO_2, TiO_2, ZrO_2, or CeO_2[4]. Within this strategy fall core-shell nanomaterials which have shown promising results in stabilizing the activity of various catalysts.[5]

Here we report the development of iron oxide nanoparticles encapsulated by SiO_2, CeO_2 and ZrO_2 for the chemical looping-based production of hydrogen. The materials were characterized by high resolution transmission electron microscope (HR-TEM), powder X-ray diffraction and N_2 adsorption. The redox performance of the synthesized oxygen carriers were assessed in a thermo-gravimetric analyser. The shell thickness is optimized to ensure a good compromise between redox stability and mass transfer characteristics. The newly developed materials show an improved stability against sintering when compared to conventionally supported oxygen carriers.

---

[1] P. Gupta, L. Velazquez-Vargas, L. Fan, Energy Fuels (2007) 21, 2900-8.
[2] R. D. Solunke and G. T. Veser, Ind. Eng. Chem. Res. (2010), 49, 11037.
[3] C. D Bohn, J. P. Cleeton, C.R. Müller, S. Y. Chuang, S. A. Scott, J. S. Dennis, Energy Fuels, (2010), 24, 4025-4033.
[4] J. Adanez, A. Abad, F. Garcia-Labiano, P. Gayan, L. F. de Diego, (2012), 38, 215-282.
[5] R. G. Chaudhuri and S. Paria, Chem. Rev. (2012), 112, 2373–2433

Fe and Ni-based Oxygen Carriers for CO_2 activation via Chemical Looping

Amey More, Saurabh Bhavsar and Götz Veser

Department of Chemical Engineering, University of Pittsburgh, Pittsburgh, PA

Abstract:

'Chemical Looping Combustion' (CLC) is an emerging clean combustion technology which offers an efficient and elegant route for fossil fuel combustion with inherent CO_2 capture based on the cyclic oxidation and reduction of an "oxygen carrier" (typically a metal) with air and a fuel, respectively. We have recently shown that this process can also be used for CO_2 activation by replacing air with CO_2 as the oxidant for the process, thus enabling the reduction of CO_2 to CO (which has many uses in the chemical industry) as an alternate pathway to carbon sequestration. We had identified Fe as the most suitable metal for the process due to its good reactivity with CO_2 at realistic process operating conditions. However, Fe shows poor reactivity with methane as a typical fuel for this process. Ni and NiO, on the other hand, is well known as an efficient catalyst for methane combustion, but cannot be oxidized with CO_2. We hence hypothesized that a bimetallic Fe/Ni oxide would show combined good reactivity with both CH_4 and CO_2 in the reduction and oxidation half cycles, respectively.

Supported Ni, Fe, and Ni/Fe oxygen carriers with 40wt% metal loading were synthesized using a simple wet-impregnation procedure. Three different ratios of the Fe to Ni were studied for their performance. Thermo-gravimetric analysis (TGA) of the carriers combined with X-Ray Diffraction (XRD) gave insight into oxidation states, carrier conversions, and reaction rates were evaluated in fixed-bed reactor experiments. We found that bimetallic oxides can indeed show activity for both methane combustion and CO_2 reduction; albeit the individual rates of oxidation are lower than for monometallic Fe carriers, and rates of reduction are lower than for monometallic Ni carriers. For comparison, physical mixtures of monometallic Fe and Ni carriers with the same metal weight loadings were prepared to isolate the oxidation characteristics of the individual metal oxides during the reactive processes. Remarkably, a significantly higher rate of carrier conversion obtained with the physical mixture—not only in comparison to the bimetallic carriers, but also in comparison to either monometallic carrier—suggested that the steps involved were quite different from those in case of a mixed oxide. The remarkable behavior can be explained based on a synergistic reaction sequence between the two carrier materials, which will be discussed in detail in the presentation. Overall, our study thus identifies a very simple, yet highly promising carrier system for an efficient chemical looping process for CO_2 activation.

383462 Continuous Regime of Clc/CLOU Reactivity and Its Impact on the Dynamic and Steady State Performance of Cu_xO_y Oxygen Carriers

Abstract:

A detailed kinetic network that captures the continuous regime of Cu-based oxygen carrier reactivity in chemical-looping combustion (CLC) and/or chemical-looping with oxygen uncoupling (CLOU) processes is used to study transient and steady state process efficiencies. Theoretical models for CLC/CLOU processes and experimental facilities are utilized for this analysis. The transient kinetics of Cu-based oxygen carriers in the temperature range of 750 to 1000 °C was investigated in a fixed-bed reactor (Figure 1), supported by supplementary characterization techniques, such as TGA, SEM and in-situ hot-stage XRD (Figure 2). The continuous regime between CLC and CLOU, as studied experimentally and theoretically is shown in Figure 3. The effect of operating pressure on the reaction kinetics was also investigated for the purpose of predicting syngas CL processes integrated in IGCC power/hydrogen plants. The established kinetic network is then implemented in different process models to evaluate system performance. The transient behavior of CLC/CLOU is simulated by a hydrodynamic three-phase (bubble, emulsion and wake) model in a fluidized bed reactor, while Aspen Plus process flowsheet simulations are performed to study the steady state behavior. On the basis of the developed Cu-based kinetics and process models, a comprehensive evaluation of the real efficiency of Cu-based CL processes integrated in the power generation infrastructure is carried out.

Acknowledgement: This material is based upon work supported by the National Science Foundation under Grant No. 1054718.

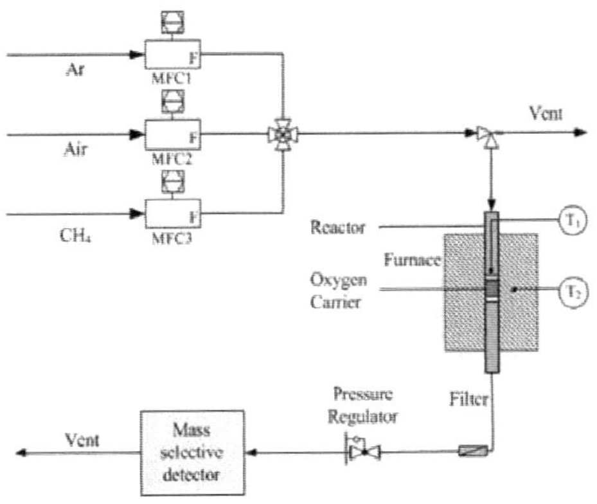

Figure 1: Fixed-bed chemical-looping setup used in this work.

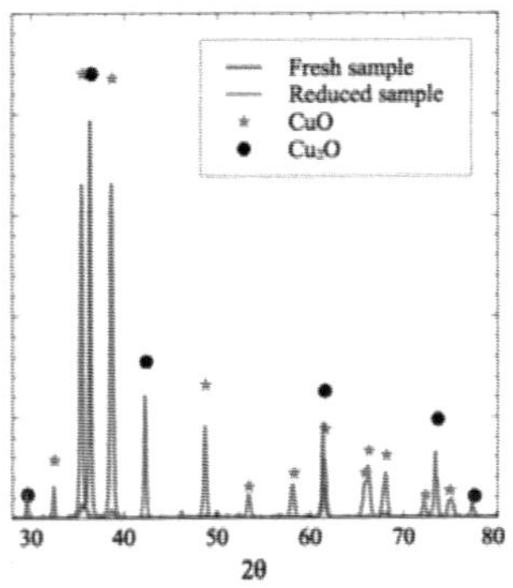

Figure 2: XRD spectra of SiO_2-supported CuO sample in oxidized and reduced (decomposed in Ar) forms.

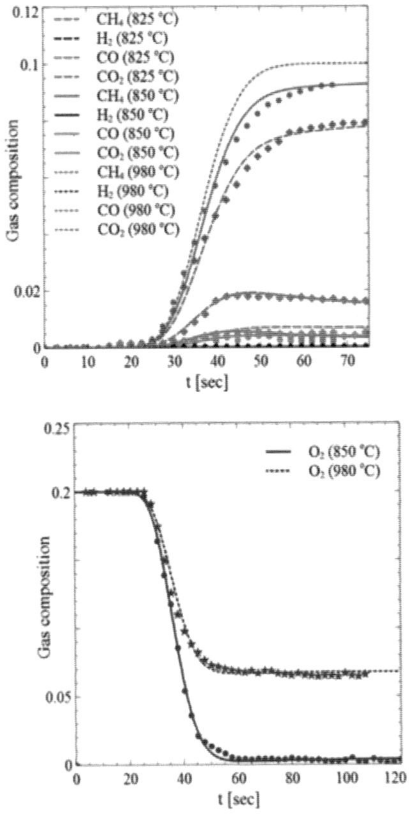

Figure 3: Model prediction and experimental data of (a) chemical-looping combustion selectivity and (b) chemical-looping with oxygen uncoupling using Cu/CuO and CH_4/Ar in a fixed-bed reactor. *Note:* model prediction at 825 °C (dashed line), 850 °C (solid line) and 980 °C (dotted line); experimental data at 825 °C (diamond), 850 °C (circle) and 980 °C (star).

Perovskite Based Oxygen Carriers for Chemical Looping with Oxygen Uncoupling (CLOU)

*Nathan Galinsky, Amit Mishra, and Fanxing Li**

Department of Chemical and Bimolecular Engineering
North Carolina State University
Raleigh, NC 27695
* Email: fli5@ncsu.edu

Chemical looping processes utilize metal oxides to store and release lattice oxygen to oxidize gaseous and solid carbonaceous fuels such as methane, syngas, coal, and biomass. To regenerate the oxygen released during the fuel oxidation step, the metal oxides, or oxygen carriers, can be regenerated by reacting with air. Typical oxygen carriers rely on Fe, Mn, Cu, Co, and/or Ni to carry lattice oxygen. An inert ceramic support is often used to enhance the activity and stability of the oxygen carrier. Perovskites have recently been shown to possess good oxygen carrying properties for different chemical looping applications including CLOU and chemical looping reforming (CLR). Such perovskite based oxygen carriers with CLOU properties allow for oxygen to be released to gas phase at high temperature, leading to improved reaction kinetics for solid fuel conversion. In this work, we investigate the effect of dopant on the CLOU and CLR characteristics of $CaMnO_3$ perovskite. A-site dopant such as Ba and B-site dopants including a number of first row transition metal oxides are investigated. While secondary metal dopants may reduce oxygen uncoupling capability of the parent perovskite by up to 50%, they are shown to be stable for over 100 redox cycles. Such effects are explained through both experiments and simulations.

In-situ X-ray absorption spectroscopy study of Na-doped and Al_2O_3-stabilized, CuO-based oxygen carriers for chemical looping combustion and chemical looping with oxygen uncoupling

Qasim IMTIAZ,[1] Paula ABDALA[2], Agnieszka KIERZKOWSKA[1], Wouter van BEEK[2], and Christoph MÜLLER[1]

[1]Laboratory of Energy Science and Engineering, ETH Zürich, Leonhardstrasse 27, 8092 Zürich, Switzerland
[2]European Synchrotron Radiation Facility (ESRF), Grenoble, France

Abstract

Chemical looping combustion (CLC) and chemical looping with oxygen uncoupling (CLOU) are emerging carbon capture and storage (CCS) technologies that are conceptually linked to oxy-fuel combustion. In CLC, hydro-carbonaceous fuels are combusted using the lattice oxygen of an oxygen carrier, typically a transition metal oxide. In the case of CLOU, the hydro-carbonaceous fuel is combusted with molecular oxygen derived from the decomposition reaction of a transition metal oxide. The CuO-Cu_2O-Cu system is very suitable for both CLC and CLOU owing to its (i) high oxygen carrying and uncoupling capacity of, respectively, 0.2 g O_2/g CuO and 0.1 g O_2/g CuO and fast kinetics for the reduction and decomposition reactions. However, CuO has one main disadvantage, *i.e.* the low Tammann temperature of CuO of 526 °C. Therefore, CuO requires stabilization by a high Tammann temperature support, *e.g.* Al_2O_3 or $MgAl_2O_4$. Al_2O_3-stabilized, CuO-based oxygen carriers have received considerable attention owing to the low cost of Al_2O_3.[i] However, Al_2O_3 can form mixed oxides with CuO, *viz.* $CuAl_2O_4$ (spinel) and $CuAlO_2$ (delafossite). Compared to the CuO-Cu_2O couple, the $CuAl_2O_4$-$CuAlO_2$ couple possesses a low oxygen equilibrium partial pressure and a slow decomposition/re-oxidation kinetics. Therefore, the formation of $CuAl_2O_4$ negatively affects the CLOU performance of Al_2O_3-stabilized, CuO-based oxygen carriers. Interestingly, the formation of the $CuAl_2O_4$ spinel can be inhibited by the incorporation of Na^+ or K^+ ions in the CuO-Al_2O_3 matrix. Here, we utilized a sol-gel technique to synthesize Na^+ promoted (1-5 wt. %) and Al_2O_3-stabilized, CuO-based oxygen carriers containing 80 wt. % CuO. The synthesized oxygen carriers were characterized using powder X-ray diffraction (XRD), X-ray absorption spectroscopy (XAS), N_2 adsorption and temperature programmed decomposition. The cyclic oxygen uncoupling capacity of the synthesized materials was evaluated in a thermo-gravimetric analyzer (TGA) at 900 °C. The oxygen uncoupling capacity of Al_2O_3-stabilized CuO decreased with cycle number owing to the slow oxidation kinetics of $CuAlO_2$. On the other hand, Na^+ promoted, CuO-based oxygen carriers showed a stable oxygen uncoupling capacity with cycle number. *In-situ* X-ray absorption spectroscopy (XAS) revealed that the presence of Na^+ ions not only inhibits the formation of $CuAl_2O_4$ but also promotes the reduction of CuO and Cu_2O. Owing to its high and stable oxygen uncoupling capacity and fast kinetcis, Na^+ doped and Al_2O_3-stabilized CuO was identified as a very promising oxygen carrier for CLC and CLOU.

---

[i] J. Adanez, A. Abad, F. Garcia-Labiano, P. Gayan, L. F. de Diego, Prog. Energy Combust. Sci. 2012, 38, 215 – 282.

DEVELOPMENT OF SOLIDS FLOW SENSOR TECHNOLOGY FOR CHEMICAL LOOPING

Benjamin T. Chorpening, U. S. Department of Energy, National Energy Technology Laboratory, Morgantown, WV, USA
David W. Greve, Carnegie Mellon University, Pittsburgh, PA, USA
Jared Charley, URS Washington Division, National Energy Technology Laboratory, Morgantown, WV, USA
Justin Weber, U. S. Department of Energy, National Energy Technology Laboratory, Morgantown, WV, USA
Douglas L. Straub, U. S. Department of Energy, National Energy Technology Laboratory, Morgantown, WV, USA

Abstract

Chemical looping combustion is being developed by the U.S. Department of Energy as a possible technology for efficient carbon dioxide separation for heat or power generation. Control of them chemical looping process, however, has proven to be a technology challenge. In order to control the chemical looping process, accurate and rapid measurement of key process parameters is required. A high temperature solids-flow sensor, based on microwave Doppler technology, is being developed to provide a real-time measurement of solids circulation rate. Initial testing has shown the feasibility of the approach. Testing of the prototype with high temperature solids is the next step.

Introduction

Chemical looping combustion is being developed with support of the U.S. Department of Energy because it will allow for carbon capture with less efficiency penalty than conventional power cycles with post-combustion capture [1,2]. Chemical looping uses an intermediary, such as a metal oxide, to collect oxygen from air in the air reactor. The oxidized particles are then transported and to the fuel reactor, where the oxygen carrier particles are reduced. The product gas stream contains mostly carbon dioxide and water vapor, which be readily separated for carbon capture, utilization and storage. An inert buffering gas is injected in the process between the reactors to help separate the oxidation and reduction processes.

One of the challenges for operation of a chemical looping combustion process is the measurement and control of the solids circulation rate. The circulation rate is important for controlling both heat transfer and chemical reaction processes. It has been observed to have large variations on a short time scale, in cold flow systems used to model the hydrodynamics. While the circulation rate may be estimated from pressure drops at various points in the process, this measurement approach is indirect, and has had considerable uncertainty.

While there are many methods employed for measurement of air conveyed solid materials, the operating temperature and pressure parameters used or examined for use in lab- or pilot-scale chemical looping systems make most methods very difficult to implement. Chalmers University (Sweden) [3], and Southeast University (China) [4] presently have lab-scale reactors using iron at 1000°C and nickel at 1040 °C. Proposed chemical-looping gas turbine systems have the highest pressure requirements (up to 21 bar) with temperatures as high as 1050°C for iron-based carriers [5,6]. In contrast, most solids flow measurement technology has been developed for atmospheric pressure, low temperature applications such as conveyance of food ingredients, such as grains and sugar, or raw materials such as plastic pellets.

Commercially available solid mass flow meters fall into several general categories: coriolis, impact plate, diverting or curved chute, momentum probe, differential pressure, optical, capacitance change, microwave, ultrasonic, magnetic, and electrostatic. The maximum process temperature specification identified in the review was 400°C, with most of the sensors designed for near-atmospheric pressure operation [7]. Many low temperature contact methods exist for measuring the flow of low temperature solids, typically bulk materials or foodstuffs. Many are on systems that do not require sealing from the surrounding air. Implementation for the chemical looping combustion environment of 1000°C with a pressurized system of these methods poses considerable technical challenges. Optical methods, while noncontact, suffer from difficulties with window fouling, which are likely to be severe in a chemical looping environment. Microwave technologies appeared promising for modification for higher temperature use, and have been pursued in this work.

The Chemical Looping Reactor (CLR) has been designed and constructed at NETL for the purpose of research and development of the chemical looping process. The chemical looping reactor operates at high temperature (~ 1000°C) and is pressurized. The design of the microwave antenna or launcher must take these factors into account.

Microwave Solids Flow Sensor Development

A high temperature solids flow sensor based on microwave Doppler technology is being developed for chemical looping. To meet the special requirements of this application, microwave sensing provides several advantages. First, it is a non-contact method which allows a pressure boundary. Second, it is less sensitive to the surface quality of the pressure boundary, or microwave window, than an optical (visible wavelength) method. Third, it has already been commercially implemented for slightly elevated temperatures (up to 200°C).

The system is calibrated using a rotating table feeder. Particulate material is fed through a drop tube to the surface of a small round disk (the table), near its edge. The particles dropped onto the rotating table are brushed off into a funnel. The rotation of the disk and the height of the gap between the drop tube and the disk control the particle feed rate. The particles drop from the funnel, gain velocity, and pass by the sensor under calibration. The particles are collected at the bottom of the drop tube and continuously monitored with a load cell to determine the actual mass vs time (solid flow rate) for the calibration. Within a limited range, the solids flow rate produced is very steady, helpful for calibration.

The table feeder, while producing a steady flow for calibrations, is limited to low temperature operation. Questions remain on the effect of high temperature on the microwave properties of carrier particles, and how that will affect their measurement. To address this issue, the High Temperature Solids Flow Verification apparatus has been constructed. A batch of particles, approximately 10 kg, is loaded into the HITSV apparatus and preheated with light fluidization. A slide-gate valve is opened to drop the hot particles through a heated tube past the sensor under test. The hot particles are collected in a container, whose mass is measured continuously with a load cell. The flow rate of the particles can be changed by the size of the slide-gate opening.

To quickly investigate the approach of microwave mass flow sensing, a commercial low temperature microwave Doppler sensor was purchased. It was modified from its standard commercial configuration for attempted application at the CLR operating temperatures, near

1000°C. In testing, the commercial system calibrated and performed well on the table feeder. At room temperature, on the cold flow model of the CLR, the sensor data trends still appeared reasonable, although in-place calibration was needed. At high temperatures on both the CLR and on the HITSV apparatus, the flow rate data no longer appeared reasonable. The cause of this behavior is still under investigation.

NETL, in collaboration with CMU, is developing a prototype high temperature microwave mass flow sensor specifically for the chemical looping application [8,9]. This approach allows adjustment of the launcher design, electronics, and data processing. The electronics and a first prototype microwave launcher have been tested at room temperature, and data processing has been developed [10]. The prototype high temperature microwave solids flow launcher has been assembled (Figure 1), and initially tested in operation on the CLR in summer 2014. While the sensor survived the harsh operating conditions, additional improvements to the data processing and calibration are planned.

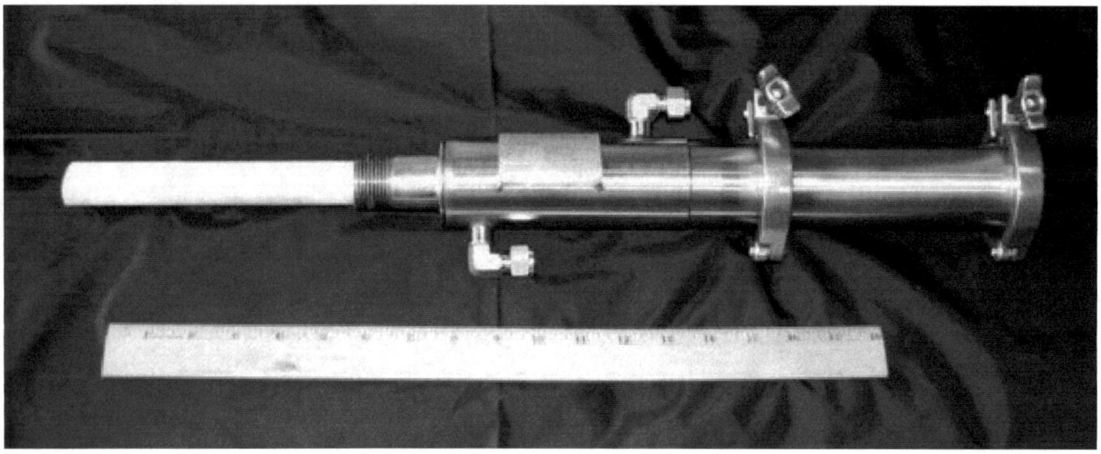

Figure 1. Prototype high temperature launcher for microwave solids flow sensor.

Acknowledgements

This project was funded by the Department of Energy, National Energy Technology Laboratory, an agency of the United States Government, through a support contract with URS Energy &Construction, Inc. Neither the United States Government nor any agency thereof, nor any of their employees, nor URS Energy & Construction, Inc., nor any of their employees, makes any warranty, expressed or implied, or assumes any legal liability or responsibility for the accuracy, completeness, or usefulness of any information, apparatus, product, or process disclosed, or represents that its use would not infringe privately owned rights. Reference herein to any specific commercial product, process, or service by trade name, trademark, manufacturer, or otherwise, does not necessarily constitute or imply its endorsement, recommendation, or favoring by the United States Government or any agency thereof. The views and opinions of authors expressed herein do not necessarily state or reflect those of the United States Government or any agency thereof.

References

1. Ekström et. al (2009), "Techno-Economic Evaluations and Benchmarking of Pre-combustion CO2 Capture and Oxy-fuel Processes Developed in the European ENCAP Project," Energy Procedia, Volume 1, Issue 1, pp. 4233-4240.
2. Alstom Power Inc., Power Plant Laboratories (2003), "Greenhouse Gas Emissions Control by Oxygen Firing in Circulating Fluidized Bed Boilers: Phase 1 – A Preliminary Systems Evaluation," Vol 1., DE-FC26-01NT41146, PPL Report PPL-03-CT-09
3. Berguerand, N., and Lyngfelt, A. (2008), "Design and Operation of a 10 kWth Chemical-Looping Combustor for Solid Fuels - Testing with South African Coal," Fuel, v. 87, pp. 2713-2726.
4. Shen, L., Wu, J. and X. Jun (2009), "Experiments on Chemical Looping Combustion of Coal with a NiO Based Oxygen Carrier," Combustion and Flame, v. 156, 721-728.
5. Lozza, Giovanni, et al. (2006), "Three Reactors Chemical Looping Combustion for High Efficiency Electricity Generation with CO2 Capture from Natural Gas," in Proceedings of ASME Turbo Expo 2006: Power for Land, Sea and Air, Barcelona, Spain, GT2006-90345.
6. Consonni, S., et al. (2004), "Chemical-Looping Combustion for Combined Cycles with CO2 Capture," in Proceedings of ASME Turbo Expo 2004: Power for Land, Sea and Air, Vienna, Austria, GT2004-53503.
7. United States, Dept. of Energy, National Energy Technology Laboratory (2011), "Evaluation of Commercially Available Solids Flow Sensors and Technologies for Chemical Looping Application," URS-RES Number URS-RES- 1-337
8. F. Gao, D. W. Greve, and I. J. Oppenheim (2012) "Launcher design for chemical looping combustion," in COMSOL Conference, Boston, Oct. 3-5.
9. F. Zhang, D. W. Greve, F. Gao, and I.J. Oppenheim (2012) "Doppler solid flow measurement for chemical looping combustion," 2nd Global Conference on Microwave Energy Applications, Long Beach, California, July 23-27.
10. Greve, D. W., Oppenheim, I. J., Chorpening, B. T. and J. Charley (2013), "Microwave Doppler flow sensor for chemical looping combustion systems," in IEEE Sensors 2013, Baltimore, Nov. 3-6.

386627 The Syngas Chemical Looping Pilot Unit Development and Demonstration

Abstract:

Chemical looping technologies have evolved into a promising alternative for the efficient conversion of carbonaceous fuels to electricity and/or high value chemicals with minimal carbon emissions. These processes utilize an oxygen carrier solid material to indirectly supply oxygen to the fuel source eliminating the need for costly and energy intensive gas-gas separation techniques. As one of the most advanced chemical looping processes to date, the Ohio State University (OSU) has successfully developed the syngas chemical looping (SCL) process for gaseous fuel conversion from small scale laboratory studies to a large scale, fully integrated pilot demonstration facility. The unique counter-current moving bed reactor design allows the OSU SCL process to achieve full fuel conversion to CO_2/H_2O while minimizing the solid circulation rate and the ability to produce high purity H_2 for further chemical production. The SCL pilot facility represents a high temperature and high pressure operation of the OSU chemical looping technology for electricity and H_2 cogeneration with nearly 100% CO_2 capture. The construction, commissioning, and preliminary operations of the SCL unit have been completed. To date, the pilot SCL unit has completed over 184 hours of integrated operation.

The present paper summarizes key developments in the SCL process used for the pilot unit design. Specifically, the thermodynamic considerations of metal oxide and reactor contacting pattern are discussed. The results from TGA and bench studies are used to develop a 1-D kinetic model of the moving bed reactor. The kinetic model results identifying the critical flow ratio of fuel to metal oxide required for the pilot unit operation are presented. An overview of the pilot SCL unit design, construction, and commissioning is provided. The operation results including solid circulation studies under high pressure and temperature conditions are also discussed.

388800 Design and Performance of a Novel Reverse-Flow Fixed-Bed Reactor for Chemical-Looping Combustion

Abstract:

Design and performance of a novel reverse-flow fixed-bed reactor for chemical-looping combustion

Lu Han, Zhiquan Zhou, George M. Bollas

Abstract

The reactor technologies for chemical-looping combustion (CLC) are categorized as circulating fluidized bed reactors, rotating reactors, and alternating flow reactors (Fig. 1). The most widely used reactor type is based on the interconnected fluidized-bed design, which has been successfully demonstrated in pilot plants built in Chalmers University of Technology (Sweden), Institute of Carboquimica (Spain), Korean Institute of Energy Research, and Vienna University of Technology to name a few. Fluidization promotes excellent contact between the gas and solids, which enables the process to achieve both high fuel conversion and CO_2 selectivity, in addition to a uniform temperature distribution, small pressure drop, and continuous operation. However, circulation of the solids requires an energy-intensive cyclone system and loop seals to prevent air/fuel slip. This harsh process makes particle attrition inevitable and creates fines that can be entrained out of the fluidization vessel. These issues can severely impact the process economics and safety of the CLC process, for example, if Ni is used as the oxygen carrier. Alternatively, CLC can be implemented in a transiently operated fixed-bed reactor where the gas flows alternate between reducing and oxidizing environments. This design eliminates the need for gas-solid separation and relieves some of the requirements of a potential oxygen carrier. Fixed-bed reactors are in general easier to operate, scale-up and pressurize and are flexible to handle a range of particle sizes. The design is also more compact than a fluidized-bed reactor, which allows for better utilization of the oxygen carrier, lower capital cost, and smaller process footprint. However, the performance of the fixed-bed process exhibits lower CO_2 selectivity, greater carbon deposition, and much higher temperature fluctuations compared to an equivalently designed fluidized bed reactor. Other proposed reactor options for CLC include the moving bed reactor, rotating reactor, and rotary reactor. These designs share in the technical difficulties that arise from putting solids in motion, specifically related to gas leakage and product gas mixing with air.

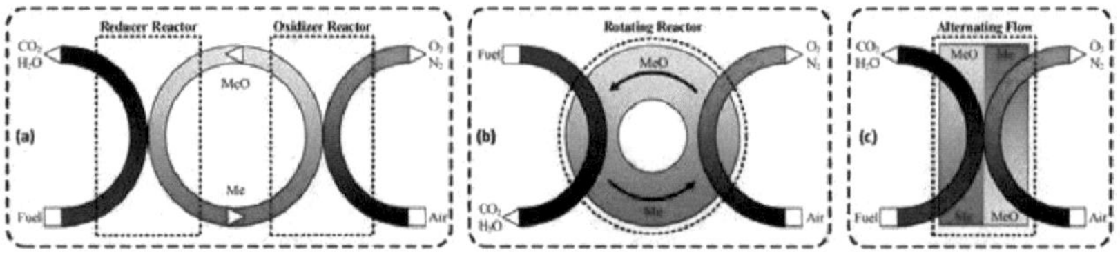

Figure 1. *Reactor options for CLC: (a) circulating fluidized-bed; (b) rotating reactor; and (c) alternating flow over a fixed-bed.*

In this paper, a novel reverse-flow reactor is proposed for CLC of gaseous fuels, that integrates gas-solid separation, reaction, and heat transfer in a single fixed-bed unit. The periodic reversal of the fuel gas during the oxygen carrier reduction cycle promotes more uniform contact between the fuel and the unconverted solids and better utilization of the conductive heat of the packed bed, while maintaining the intrinsic advantages of a stationary bed. Reverse-flow reactors had been traditionally employed as a method to improve the heat integration for many industrially relevant reactions, including SO_2 oxidation, total oxidation of hydrocarbons in off-gases, and NO_x reduction. When applied to CLC of gaseous fuels, the reverse-flow reactor achieves higher CO_2 selectivity for all practical oxygen carrier conversions compared to a traditional fixed-bed process, at no expense of the fuel conversion. Frequently reversing the direction of the flow provides a uniform conversion profile inside the bed and yields an even higher CO_2 selectivity. Observing the CO_2 capture efficiency of the process (defined as time-integral of the CO_2 product distribution over the total flow of fuel fed), the traditional one-directional reactor quickly drops below 90% at an overall bed conversion of 35% while the reverse-flow reactor can be continually operated till a conversion of 60% to meet the same efficiency. These results are comparable to a circulating fluidized bed unit, which achieves about 94% CO_2 gas yield for a solid conversion of 60%. Additional benefits to the heat transfer characteristics are also realized with the reverse-flow design, where at least a 50 reduction in temperature drop is exhibited. The results of this work demonstrate that periodic reversal of the gas flow during CLC reduction leads to significant improvements in reactor performance, in terms of CO_2 selectivity, CO_2 capture efficiency, and bed temperature fluctuations. This fixed-bed design avoids the issue of attrition and gas-solid separation commonly found in circulating fluidized-bed units. Furthermore, the simplicity of the process makes it easily applicable to existing bench-scale units without considerably complicated equipment, for which scale-up does not entail significant challenges. Further design considerations and areas for optimization will be discussed.

Acknowledgement: This material is based upon work supported by the National Science Foundation under Grant No. 1054718.

390298 Characterization of Fuel Conversion in Coal Char-Fueled Chemical Looping Combustion Using Iron Based Oxygen Carrier

Abstract:

Chemical looping combustion (CLC) for power generation has the inherent potential of enriching CO_2 from flue gases with low energy penalty compared to other commercial available options. CLC process usually includes two reactors, an air and fuel reactor. Metal-based cycling materials (named oxygen carrier, OC) is used to transport oxygen from air reactor to fuel reactor where the fuel were burned via reduction of oxygen carrier, and thus ASU (Air Separation Unit) process is vitally avoided. When the application of CLC to solid fuel, solid fuel would indirectly reacts with oxygen carrier via steam or CO_2gasification. The fuel conversion in the fuel reactor would involve more reactions than a gas fueled CLC- coal char gasification to produce syngas and the intermediate syngas combustion with oxygen provided by the cycling materials. However, the overall reaction rate and the performance of solid-fueled CLC process are limited by fuel conversion. The purpose of this study is to characterize the char gasification and intermediate syngas conversion in char-fueled CLC process and to try to address the efficient pathway for improvement.

A series of experiments were performance on a bench-scale fluidized bed reactor to investigate solid-fueled CLC. Steam and iron-based cycling materials prepared from industrial solid waste were used as gasification agent and oxygen donator respectively.

The first batch of experiments were carried out to identify the rate-limiting step as well as the potential of improvement in the overall reaction taking place in the fuel reactor. Five different experiments were carried out on the fluidized bed reactor, including (1) coal char combustion; (2) coal char pyrolysis; (3) steam-char gasification with quartz sands as bed materials; (4) coal char fueled CLC with iron-based materials as OCs, (5) solid-solid reaction among coal char and OC, and (6) OC reduction with 20 vol% CO. It is confirmed the in-situ gasification reaction of coal char is the dominant, time-limiting factor in CLC, but in CLC process the iron-based OC can quickly remove the syngas from char gasification and shows the potential to promote the overall reaction rate of solid fueled CLC to the same level of char combustion if in-situ gasification is not the rate-limiting step.

The second batch of experiments was a parametric study. The effects of volume concentration of gasification agent, operation temperature, different coal char and fuel char/OC ratio were extensively investigated. For comparison, steam-char gasification operated under the same conditions was investigated and the resulting carbon conversion rate was used as a baseline to evaluate the performance of CLC.

This study also tries to obtain kinetics of the coal char gasification in CLC (still ongoing). Because the oxygen carrier affects the local concentration of reactant and product gas, and the structure of a fuel char particle will change with conversion, the initial rate of gasification will be derived for kinetic study.

14AIChE
Annual Meeting, Atlanta, GA

Chemical Looping Hydrogen Production by the Steam Iron Process

Viktor Hacker, Stephan Nestl, Gernot Voitic

Institute of Chemical Engineering and Environmental Technologies, NAWI Graz
Graz University of Technology

8010 Graz, Inffeldgasse 25C, Austria
viktor.hacker@tugraz.at

Content

- Production of pure hydrogen
- The concept of the reformer-steam iron process
- High pressure hydrogen production
- Prototype development
- Summary

Hydrogen production

Decentralised hydrogen production suffers from several problems:

- Low overall efficiency
- Lack of small-scale purification systems
- Safe and stable hydrogen storage technologies
- Hydrogen compression efficiency

The reformer-steam iron process

The **combination of conventional steam reforming** with the **steam iron process** in a **fixed bed reactor** enables the **decentralised production of pure hydrogen** in one simple and compact unit.

$CH_4 + H_2O \rightarrow CO + 3H_2$

$Fe_2O_3 + 3H_2/CO \rightarrow 2Fe + H_2O/CO_2$

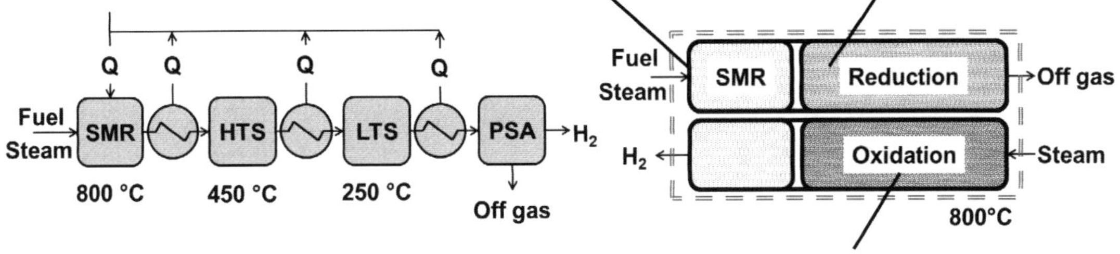

$3Fe + 4H_2O \rightarrow Fe_3O_4 + 4H_2$

Hacker, V.: A novel process for stationary hydrogen production: the reformer sponge iron cycle (RESC). J. of power sources 118 (2003) 1-2, S. 311 - 314

The reformer-steam iron process

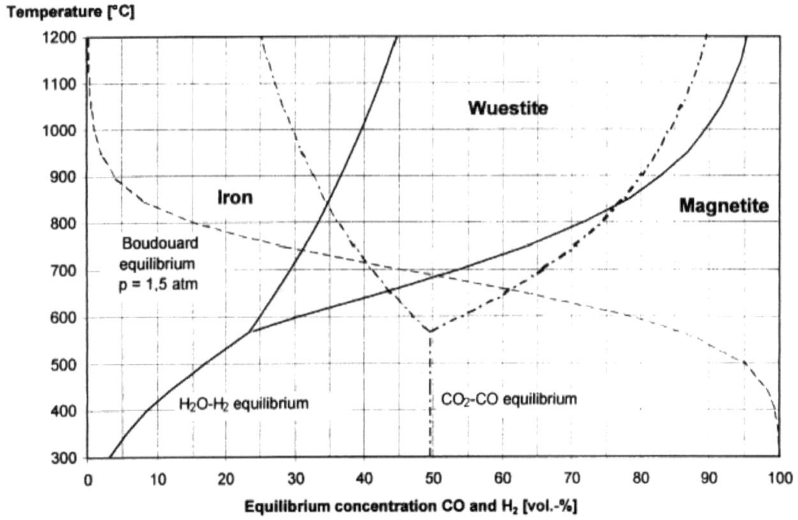

Baur-Glaessner-diagramm

Fraser, S.; Monsberger, M.; Hacker, V.: A thermodynamic analysis of the Reformer Sponge Iron Cycle. J. power sources 161 (2006) 1, S. 420 - 431

Research Studio Austria: Decentralised HYDROGEN Production - The Flex Fuel Reformer

The RESC Reformer-Steam-Iron-Cycle

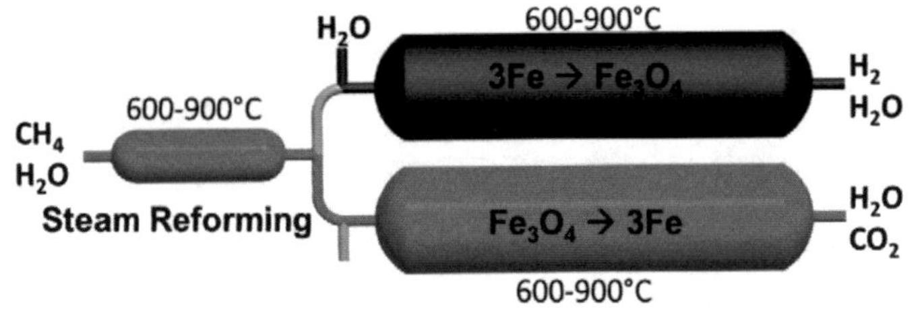

Prototyp reactor to produce and purify H_2 based on the Reformer-Steam-Iron-Cycle

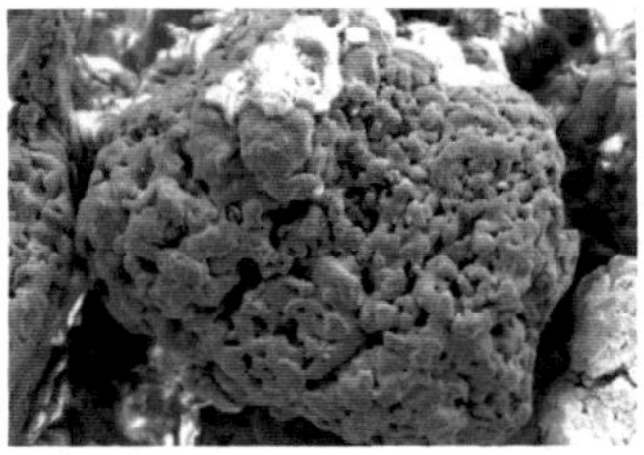

REM-EDX Image of an iron particle

bmwfi FFG

The reformer-steam iron process

- Catalyst: Ni
- Oxygen carrier: Fe_2O_3/Al_2O_3
- CH_4-flow: 300 ml·min^{-1}
- Steam to CH_4 ratio: 3.6
- Temperature: 800°C
- Oxidation: 1.4 g·min^{-1} steam
- N_2-flow: 100 ml·min^{-1}

The reformer-steam iron process

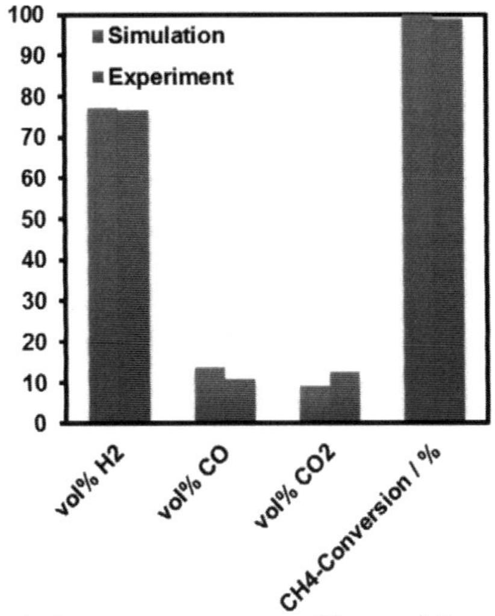

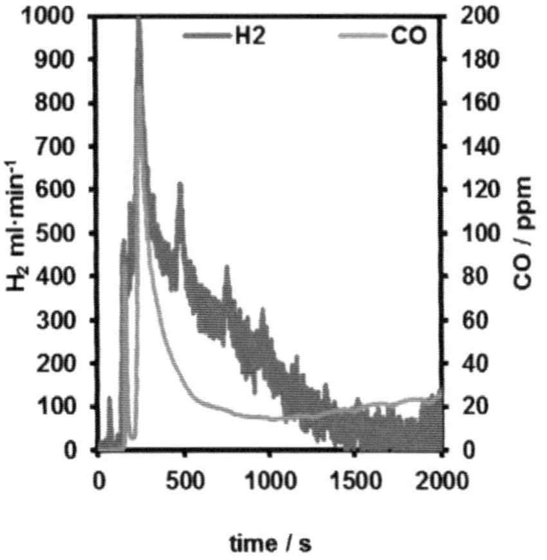

Measured dry gas composition of the reformer off gas compared with **calculated gas composition** using HSC Chemistry® (left) and **hydrogen production and CO content** (right).

High pressure hydrogen production

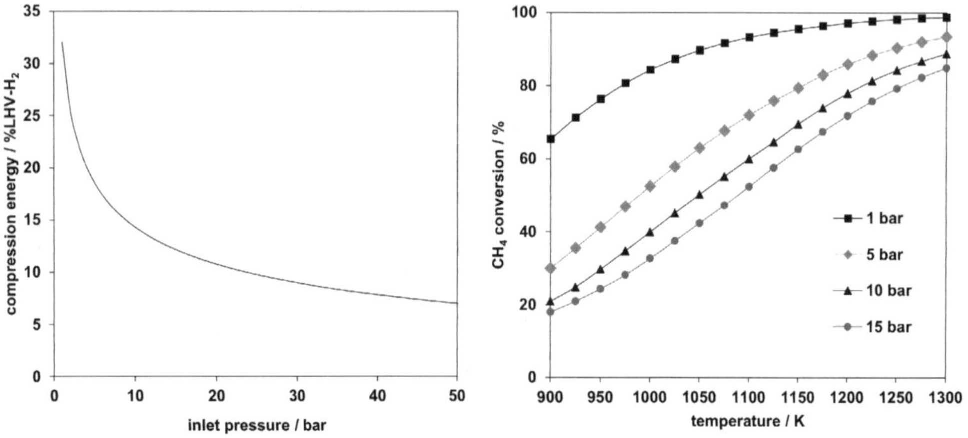

Energy demand for the **compression of hydrogen** as a function of the inlet pressure (left) and the **influence of pressure** on the **CH_4 conversion** at steam reforming (right).

High pressure hydrogen production

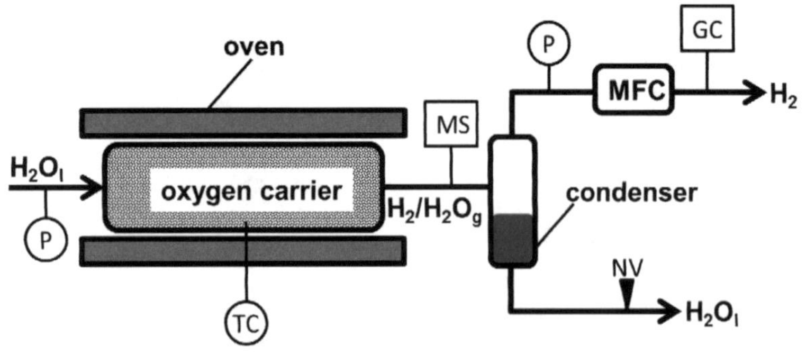

- P, TC... pressure and temperature measurement
- NV... needle valve
- MFC... mass flow controller
- GC/MS... gas analysis

High pressure hydrogen production

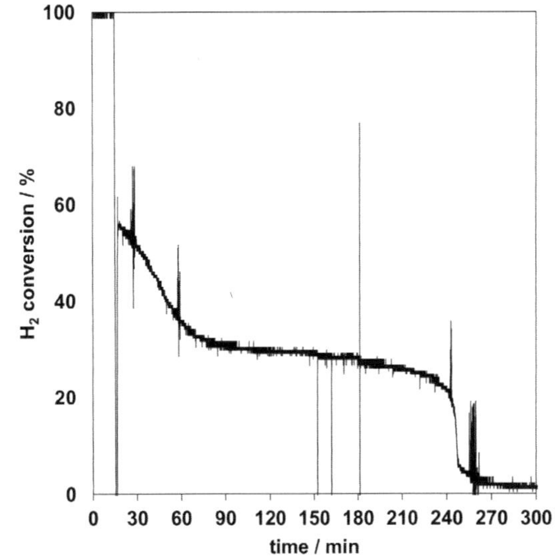

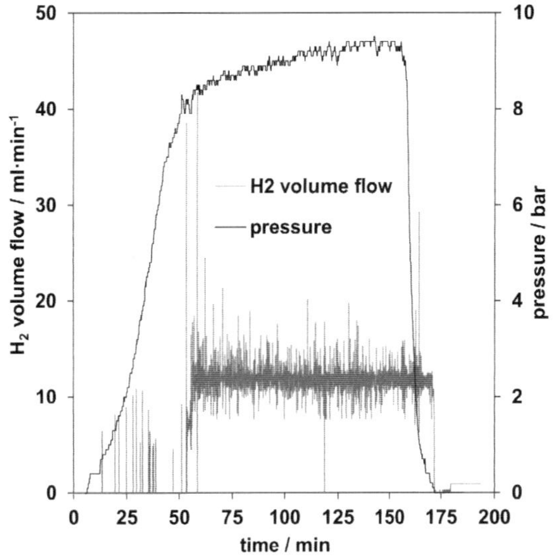

Prototype development

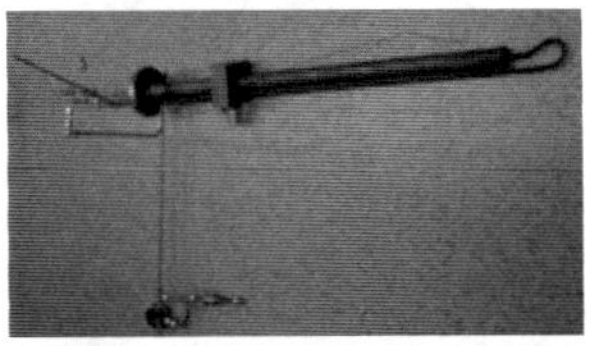

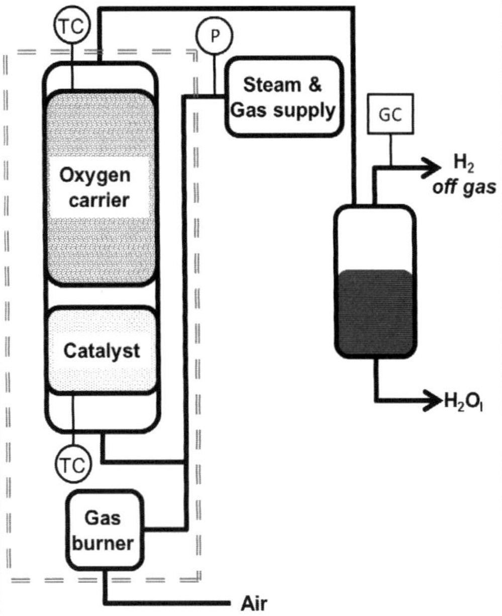

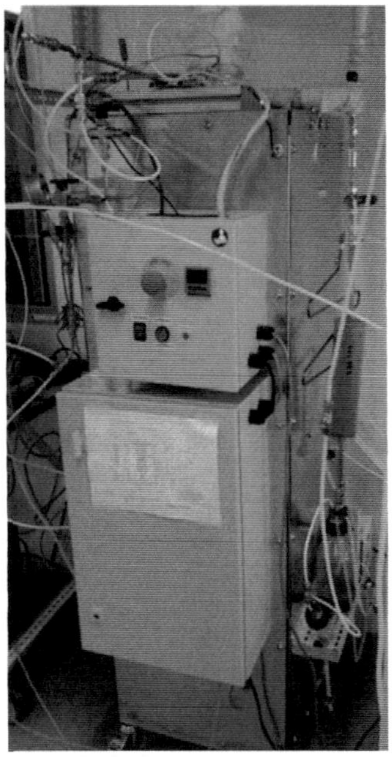

Prototype development

Temperature profile of a reduction-oxidation cycle

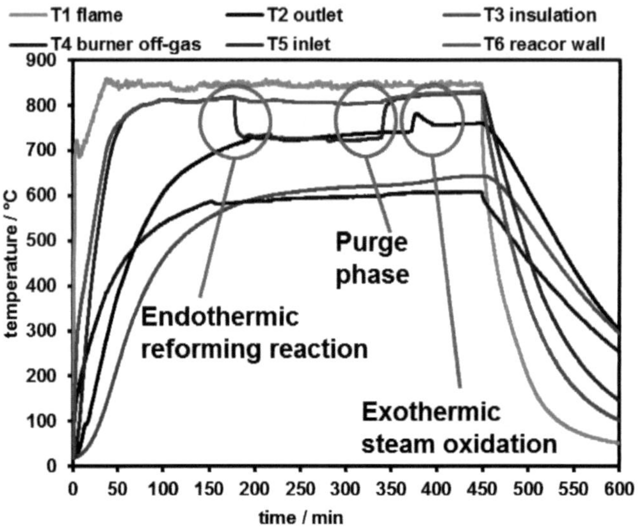

Prototype development

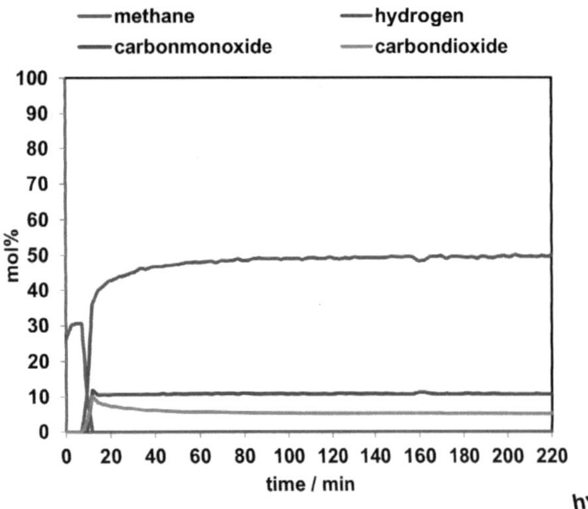

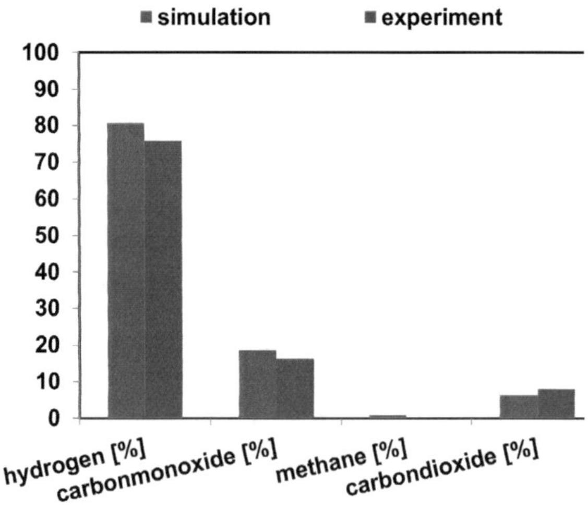

S/C ratio = 2 (CH_4 560 ml·min^{-1}); reforming temperatur: 730 °C; reduction and oxidation temperatur 750 °C

Prototype development

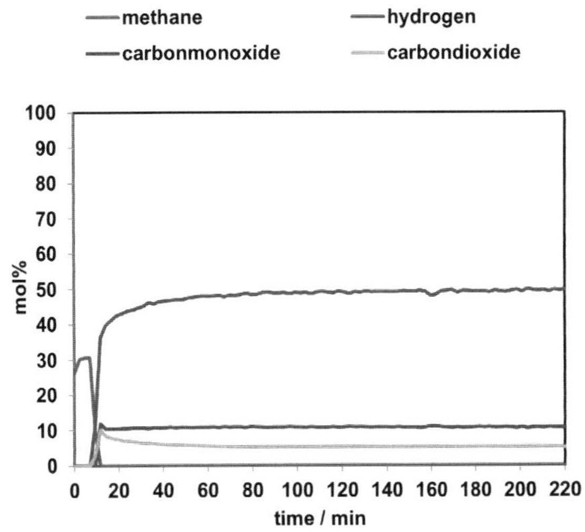

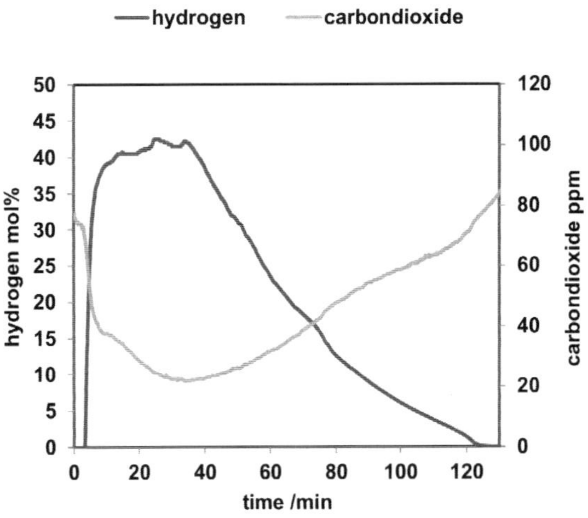

Summary

- **High-purity hydrogen** was produced from syngas, biogas and CH_4 using **fixed bed reactor systems**
- The **production of** pure **pressurised hydrogen** is possible **without additional gas compression** by steam oxidation of iron
- A **1 kW-prototype** for the **production of pure hydrogen** using **CH_4 or biogas as fuel** is under development

2014 AIChE Annual Meeting

TU Graz

Acknowledgments

Thank you for your attention!

FFG

NAWI Graz

bmwfw
Federal Ministry of Science, Research and Economy

17 Institute of Chemical Engineering and Environmental Technology 20.11.2014 CEET

369153 Biomass Gasification/Combustion Using NiO As an Oxygen Carrier in Clc Process

Abstract:

Heavier hydrocarbon fuels such as petroleum residue (liquid), solid fuel such as coal and biomass is considerably abundant and cheaper than natural gas. Considering the facts several research groups recently being devoted their efforts to adapt solid fuels such as petroleum coke, coal and biomass in CLC. This process considers in situ gasification of solid fuel to produce and burn the produced syngas inside the fuel reactor of a CLC process. In this option the solid-solid reaction between the solid fuel and the solid oxygen carrier is likely to be the rate limiting step. The agricultural and municipal wastes have long been considered as potential major source of renewable energy with zero net CO_2 emissions. The use of biomass as fuel, in conjunction with CLC, offers the additional possibility of delivering negative CO_2emissions. To accomplish this, a possible approach is to gasify the solid fuel producing syngas which can be subsequently burned inside the same fuel reactor of a CLC process. This approach has the following main advantages: (i) eliminates extra units for solid gasification and (ii) circumvents the dilution of the syngas with nitrogen.

This communication reports the investigation of integrated biomass gasification and chemical-looping combustion (CLC) process for CO_2 capture from a biomass based power generating stations. In this regard, a lanthanum modified Ni/La-γAl_2O_3 material is prepared and studied as an oxygen carrier. Temperature programmed reduction characterization of Ni/La-γAl_2O_3 shows that the presence of La enhances the reducibility of the oxygen carrier by influencing the metal-support interactions helping the formation of reducible nickel species. The integrated biomass gasification and CLC experiments are developed in a CREC Fluidized Riser Simulator using glucose as model compound for biomass. The product analysis shows that CO_2 and H_2O as major products indicating that the combustion of the biomass gasification products can be achieved using an oxygen carrier materials.

Keywords: Biomass gasification, CO_2capture, NiO oxygen carrier.

Acknowledgement: We would like to gratefully acknowledge the support provided by King Abdulaziz City for Science and Technology (KACST) through KACST-TIC on CCS at KFUPM.

370954 Economic Feasibility Analysis of Coal-Direct Chemical Looping

Abstract:

Coal-direct chemical looping (CDCL) is an advanced oxy-combustion technology that has the potential to enable substantial savings in the cost and energy penalty associated with carbon dioxide (CO_2) capture from coal-fired power plants. The CDCL utilizes an iron (III)-based oxygen carrier (Fe_2O_3) to supply oxygen for coal combustion. The process consists of a unique moving bed reactor, namely the reducer, where pulverized or crushed coal is fully converted with the reduction of iron-based oxygen carrier particles. The oxygen carrier is reduced from Fe_2O_3 to a mixture of FeO and Fe, while the flue gas leaving the reducer is a stream mainly composed of CO_2 that can be sequestered. The oxygen carrier particles contain reduced iron oxide (FeO and Fe) which are then re-oxidized to Fe_2O_3 in the combustor with air, liberating heat to produce steam for power generation.

The iron-based CDCL process has shown the potential for lower capital and operating costs as compared to first generation carbon capture technologies, such as an amine-based solvent system or traditional oxy-combustion system. Further, CDCL does not require an air separation unit (ASU). Eliminating the ASU results in a significant reduction in capital and operating costs. Through collaborative efforts, Babcock & Wilcox Power Generation Group, Inc. (B&W PGG) and The Ohio State University (OSU) have developed a preliminary design and operating philosophy for a 550 MWe commercial scale CDCL power plant. Based on the results of a techno-economic evaluation, B&W PGG estimates that the CDCL process will achieve 96.5% CO_2 capture with only a 26.8% increase in the cost of electricity (COE) when compared to a supercritical pulverized coal-fired power plant. Results of this the techno-economic analysis exceed the U.S. Department of Energy's (DOE) goal of 90% CO_2 capture at a less than 35% increase in COE. The preliminary design, the latest experimental data, and results from the techno-economic study are presented and discussed.

378018 Development of the Ohio State Natural-Gas Conversion Technology Using Iron-Based Chemical Looping Process for Hydrogen and Electricity Production

Abstract:

The global energy consumption is projected to rise by 36% in the year 2035, with fossil fuels like natural gas expected to supply for around 53% of the increase. The importance of natural gas utilization for production of electricity and chemical intermediates such as hydrogen and syngas is highlighted by the current utilization estimates, i.e.,23 trillion cubic feet, and the projected increase in consumption by 44% in 2035. The recent technological developments in natural gas utilization focus on minimizing the cost increment incurred by reduction in carbon emissions. The chemical looping technology in general is projected as a promising candidate to give the maximum cost-benefit in a carbon constrained scenario for natural gas utilization.

The syngas chemical looping technology developed at The Ohio State University employs the iron-based oxygen carrier and the gas-solid countercurrent moving-bed reactoras a reducer. This technology has been demonstrated at 25 kW_{th} sub-pilot scales using various feedstocks for near 1000 hours. This presentation analyzes the natural gas conversion effect using this technology. Specifically, the study examines the thermodynamic limits for full conversion of natural gas and the reduction kinetics under high pressure conditions for oxygen carrier reactions with methane. The results of this study are used to delineate the overall cold-gas and thermal efficiency for H_2 and electricity generation for the syngas chemical looping process system. The comprehensive techno-economic analysis based on the process efficiency and their critical comparisons with the existing technologies such as natural gas combined cycle and steam-methane reforming are also made.

393148 INVESTIGATION OF HYDRODYNAMICS FOR ALSTOM'S LIMESTONE-BASED CHEMICAL LOOPING PROCESS

Abstract:

Chemical looping is a clean-coal technology which utilizes a solid oxygen carrier to provide oxygen to a fuel conversion (combustion and gasification) process. This breakthrough fuel conversion technology does not require a capital-intensive high power consuming cryogenic air separation unit and is "transformational" in terms of its overall efficiency and cost. Chemical looping combustion for steam power is the lowest cost technical approach that ALSTOM has identified to date for coal power with carbon capture.

ALSTOM is currently developing a Limestone-based Chemical Looping Combustion (LCL-C™) technology for power generation utilizing coal. The limestone chemical looping concept is now in the advanced stages of testing at the 3-MW_{TH} prototype facility in Windsor, CT; after achieving over 350 hours of operation including 75 hours of self-sustained autothermal operation. Testing continues at the 3-MW_{TH} pilot test facility in Windsor, CT; with the goal of addressing technology gaps identified so far, and developing information necessary for further scale-up to 10 to 25 MWe. ALSTOM is concurrently developing Limestone-based Chemical Looping Gasification (LCL-G™), and Hydrogen (LCL-H™) by adaptation of LCL-C™.

As part of the development effort, Cold Flow Modeling (CFM) and Computational Fluid Dynamic (CFD) simulations are being performed to better understand the hydrodynamic behavior of oxygen carrier particles. The current focus is in evaluating the impact of product gas generation in the reducer (fuel reactor) loop on the performance of the cyclone and the dipleg.

ALSTOM utilized its licensed Barracuda® software to create a CFD model of the 3 MW_{TH} LCL-C™ pilot facilities. The CFD model was used to create a baseline comparison to a 3 MW_{TH} prototype cold flow test at ambient conditions. Next, the baseline CFD model was adjusted to simulate gas generation in the dipleg. Gas generation was simulated in an attempt to represent the instabilities resulting from the chemical looping reaction mechanism. In parallel, ALSTOM utilized their 40 foot tall plexiglass CFM to run similar tests characterizing the phenomena associated with gas generation in the dipleg. The 40 foot CFM was used to assess possible equipment, or operational modifications to provide solutions to the instabilities. Results of the CFM, and CFD work are used as a tool to gain insight into the hydrodynamics of the 3 MW_{TH} LCL-C™ prototype. This paper presents the simulation results and cold flow modeling data that were generated in association with the activities described above.

387041 Modeling and Simulation of Autothermal Chemical Looping Reforming

Abstract:

Steam Methane Reforming (SMR) is widely used for the mass production of synthesis gas and hydrogen. The heat required for the strongly endothermic SMR reactions is provided through the combustion of a fuel in a furnace in which the packed bed SMR reactor tubes are suspended. The CO2 produced in the furnace is mixed with N2 and, hence, challenging to capture. To respond to growing concerns on CO2 emissions and the energy cost of SMR, autothermal Chemical Looping Reforming technology (a-CLR) has been recently developed [1-3]. A circulating fluidized bed system is proposed. The catalyst serves for the transport of heat and oxygen between the catalyst re-oxidation reactor and the SMR reactor. In the latter, the catalyst is reduced and heat is net consumed. Direct contact between fuel and air is avoided. Because of the use of fluidized bed technology, operation at low pressure is preferred and smaller catalyst particles can be used. This allows avoiding the strong intra-particle diffusion limitations faced in the conventional packed bed SMR reactors.

Studies on a-CLR focus on lab- and pilot-scale demonstration [4-10] and on the influence of different operating variables [11]. The reaction mechanism is typically described in terms of global reactions only. A key issue for the scale-up of a-CLR technology is the development of a simulation model that accounts for the details of the reaction mechanism and kinetics. The modeling and simulation of a-CLR on a Ni-MgAl2O4 catalyst is addressed in this work. Based on kinetic modeling studies on methane steam reforming [12-19], the partial oxidation of methane [20-22], and the re-oxidation of the catalyst [23], a detailed reaction mechanism is constructed and the reaction kinetics derived. The rate equations account for the balance between oxidized and non-oxidized metal catalyst sites, oxidized sites not being active for the steam reforming and water gas shift reactions. Different reaction steps for the reduction of the metal oxide by hydrogen or methane are considered. A heterogeneous reactor model is used and validated by means of lab- and pilot scale data from the literature [9,10]. Next, commercial scale a-CLR is considered and the influence of the operating conditions studied. The conversions that can be achieved and the eventual overheating of the catalyst and related risk of sintering during the catalyst re-oxidation are evaluated.

References:

[1] Ortiz M, Abad A, de Diego LF, García-Labiano F, Gayán P, Adánez J, Int. J. Hydrogen Energ., 36, 9663-9672, 2011.

[2] Moghtaderi B, Energy Fuels, 26, 15-40, 2012.

[3] Adanez J, Abad A, Garcia-Labiano F, Gayan P, de Diego LF, Prog. Energ. Combust., 38, 215-282, 2012.

[4] Rydén M, Lyngfelt A, Mattisson T, Fuel, 85, 1631-1641, 2006.

[5] Rydén M, Lyngfelt A, Mattisson T, Energy & Fuels, 22, 2585-2597, 2008.

[6] Rydén M, Johansson M, Lyngfelt A, Mattisson T, Energy Environ. Sci., 2, 970-981, 2009.

[7] de Diego LF, Ortiz M, García-Labiano F, Adánez J, Abad A, Gayán P, J. Power Sources, 192, 27-34, 2009.

[8] Pröll T, Bolhàr-Nordenkampf J, Kolbitsch P, Hofbauer H, Fuel, 89, 1249-1256, 2010.

[9] Linderholm C, Mattisson T, Lyngfelt A, Fuel, 88, 2083-2096, 2009.

[10] Kronberger B, Lyngfelt A, Löffler G, Hofbauer H, Ind. Eng. Chem. Res., 44, 546-556, 2005.

[11] Ortiz M, de Diego LF, Abad A, García-Labiano F, Gayán P, Adánez J, Int. J. Hydrogen Energ., 35, 151-160, 2010.

[12] Xu J, Froment GF, AIChE J., 35, 88-96, 1989.

[13] Snoeck J-W, Froment GF, Ind. Eng. Chem. Res., 41, 3548-3556, 2002.

[14] Snoeck J-W, Froment GF, Ind. Eng. Chem. Res., 41, 4252-4265, 2002.

[15] Wei J, Iglesia E, J. Catal., 224, 370-383, 2004.

[16] Wei J, Iglesia E, J. Catal., 225, 116-127, 2004.

[17] Wei J, Iglesia E, J. Phys. Chem. B, 108, 4094-4103, 2004.

[18] Jones G, Jakobsen JG, Shim SS, Kleis J, Andersson MP, Rossmeisl J, Abild-Pedersen F, Bligaard T, Helveg S, Hinnemann B, Rostrup-Nielsen JR, Chorkendorff I, Shested J, Nørskov JK, J. Catal., 259, 147-160, 2008.

[19] Blaylock DW, Ogura T, Green WH, Beran GJO, J. Phys. Chem. C, 113, 4898-4908, 2009.

[20] De Groote AM, Froment GF, Appl. Catal. A: General, 138, 245-264, 1996.

[21] Slaa JC, Berger RJ, Marin GB, Catal. Let., 43, 63-70, 1997.

[22] Trim DL, Lam C-W, Chem. Eng. Sci., 35, 1405-1413, 1980.

[23] Hatcher WJ, Viville L, Froment GF, Ind. Eng. Chem. Process Des. Dev., 17, 491, 1978.

Thermal Efficiency of Combined Cycle Power Plants Involving Chemical Looping Based Systems

Vicente Rico-Ramirez (with Moises Petriz-Prieto and Guillermo Gonzalez-Alatorre)

Instituto Tecnologico de Celaya, Mexico

14AIChE
Annual Meeting, Atlanta, GA

Instituto Tecnologico de Celaya Guanajuato, Mexico

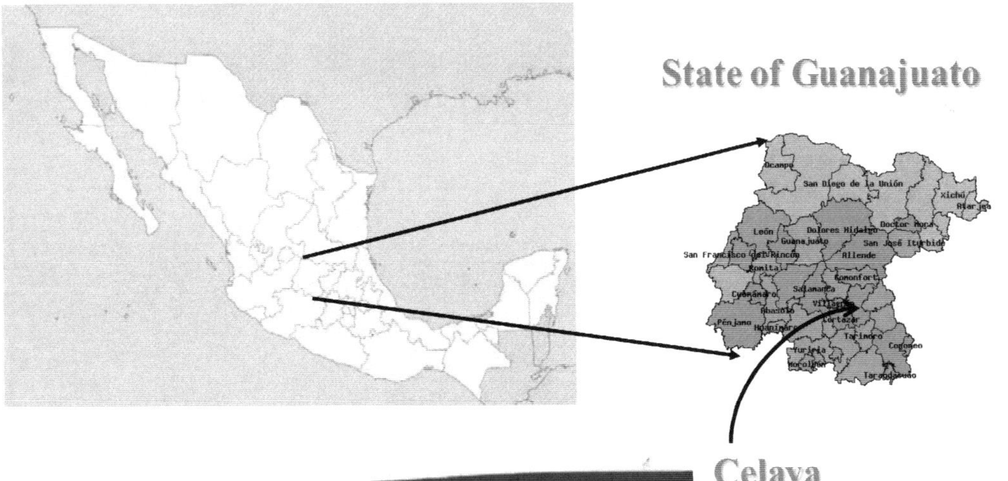

14AIChE
Annual Meeting, Atlanta, GA

- ***Basic Idea***

Chemical Looping Combustion integrated to Power Generation Configurations

Simulation Exercise: Economic Assessment, net LHV plant efficiencies

- **CLC, CLCex, CLC3**
- **SC, HAT, STIG**
- **Carrier, CLC reactors pressure, TIT**

- ***Today: Share and Discuss Our Results***

14AIChE Annual Meeting, Atlanta, GA

Outline

1 Introduction

1.1 Motivation

1.2 Chemical Looping Systems and Power Generation Cycles

2 Simulation Setup

2.1 Configurations

2.2 Models and Modeling Assumptions

2.3 Additional Considerations

3 Results

3.1 Net LHV Plant efficiencies

3.2 Economic Assessment

4 Discussion and Final Remarks

14AIChE Annual Meeting, Atlanta, GA

1 Introduction

Motivation

14AIChE
Annual Meeting, Atlanta, GA

Motivation: Known Facts

- Electric power plants using fossil fuels produce **more than one third** of the global carbon emissions

% CO_2 Emissions from Fossil Fuels (USEPA, 2011)

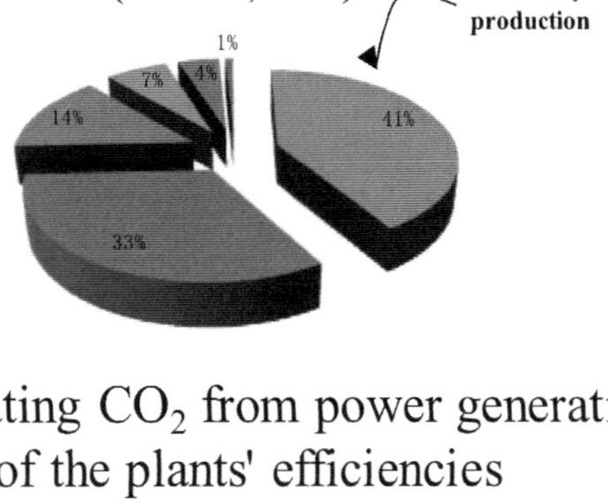

- Approaches to mitigating CO_2 from power generation include
  1 The improvement of the plants' efficiencies
  2 Increased use of renewable energy sources
  3 The capture of CO_2 emissions from the energy system
  4 Use of low-carbon fuels such as natural gas

14AIChE
Annual Meeting, Atlanta, GA

Motivation: Known Facts

- CO_2 capture in a natural gas power plant based on state-of-the-art technologies reduces the thermal efficiency by about 10%
- **Chemical Looping Combustion** (CLC) has the potential to separate CO_2 from a power generation process **with small penalties in efficiency**

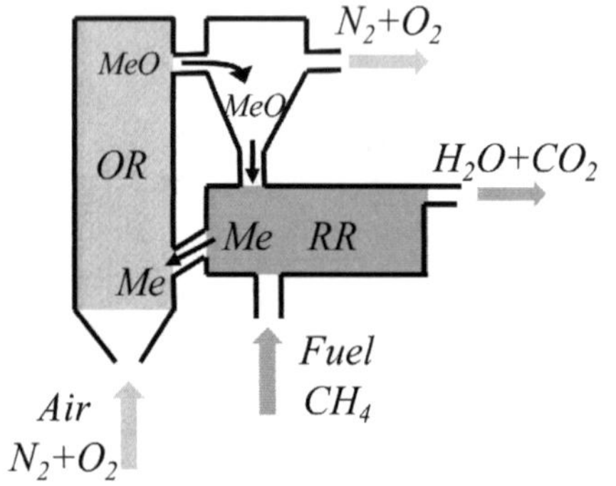

- Direct contact of fuel and combustion air is avoided through a oxygen carrier
- CO_2 is inherently separated from nitrogen

14AIChE
Annual Meeting, Atlanta, GA

Motivation: Known Facts

- Natural gas-fired combined cycles (NGCC) achieve electrical efficiencies up to 60%
- Considering a penalty of 10% due to CO_2 capture and compression (MEA absorption), these processes still can reach an efficiency of about 50%
- Therefore, to compete with commercially available technologies, a power plant with CLC has to achieve net plant LHV efficiencies of 50% or higher

↓

Study for net plant efficiency and economics assessment

14AIChE
Annual Meeting, Atlanta, GA

Chemical Looping Configurations

14AIChE
Annual Meeting, Atlanta, GA

CLC Configurations

- *Simple CLC*
- *Extended CLC, exCLC*
- *CLC 3*

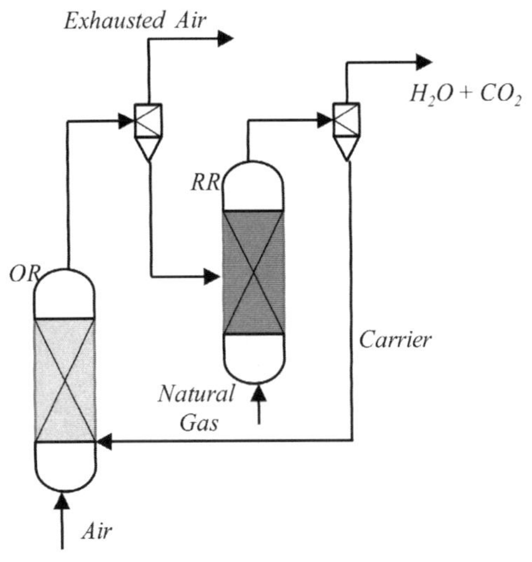

Simple CLC

- *Oxidation and Reduction reactors*
- *Inlet: Air , Fuel and Oxygen Carrier*
- *Outlet: Exhausted air, CO_2 and H_2O*

14AIChE
Annual Meeting, Atlanta, GA

Extended CLC

Extended CLC, exCLC (Wolf and Yan, 2005)

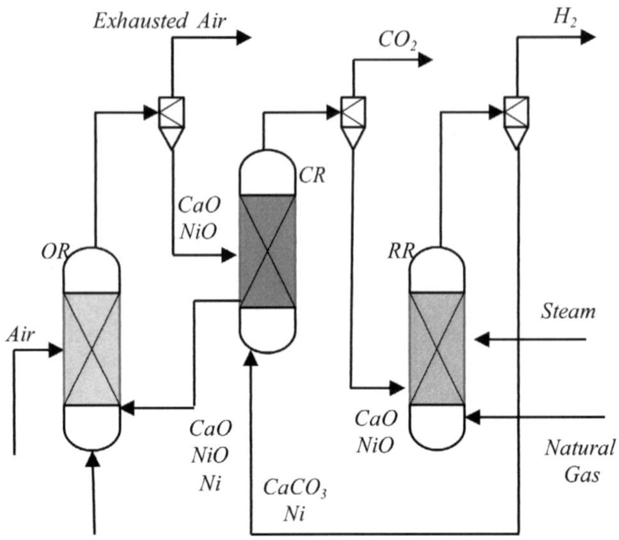

exCLC

- ***Hydrogen is produced and inherently separated from CO_2 and exhausted air***
- ***Oxidation, Reduction and Calcination Reactors***
- ***Inlet: Air , Fuel and Oxygen, Carbon Carrier***
- ***Outlet: Exhausted air, CO_2, H_2 and H_2O***

CLC3

CLC3 (Chiesa et al, 2008)

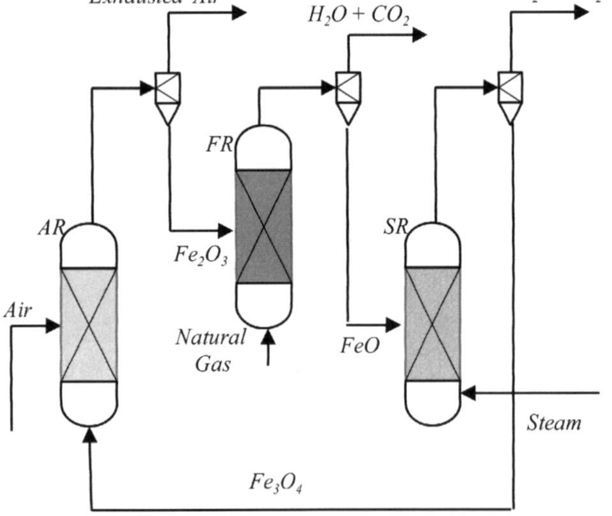

CLC3

- ***Hydrogen is also produced and inherently separated from CO_2***
- ***Fuel Reactor, Steam Reactor and Air Reactor***
- ***Inlet: Air , Steam, Fuel and Oxygen Carrier***
- ***Outlet: Exhausted air, CO_2, H_2 and H_2O***

Power Generation Cycles and Plants Under Study

14AIChE
Annual Meeting, Atlanta, GA

Natural Gas Combined Cycle

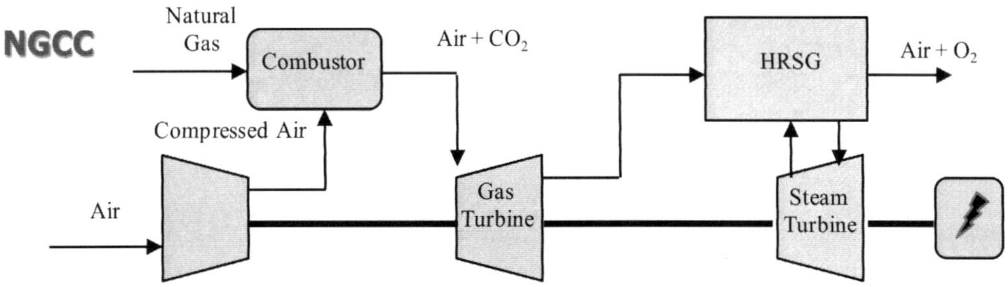

- *Conventional commercially available components*
- *Gas Turbine + Steam Turbine*

14AIChE
Annual Meeting, Atlanta, GA

Steam Injected Gas Turbine Cycle (STIG)

STIG

(Steam Injected Gas Turbine)

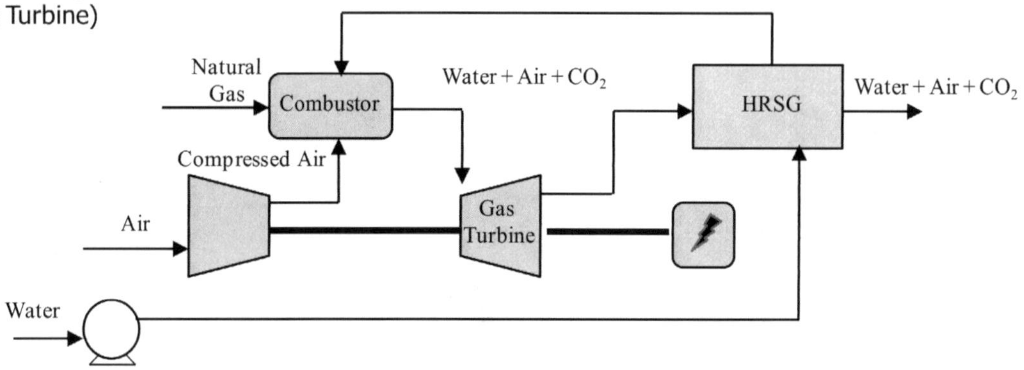

- ***Water injected into the combustion chamber increases Turbine work (by increasing working fluid)***

14AIChE Annual Meeting, Atlanta, GA

HAT: Humid Air Turbine

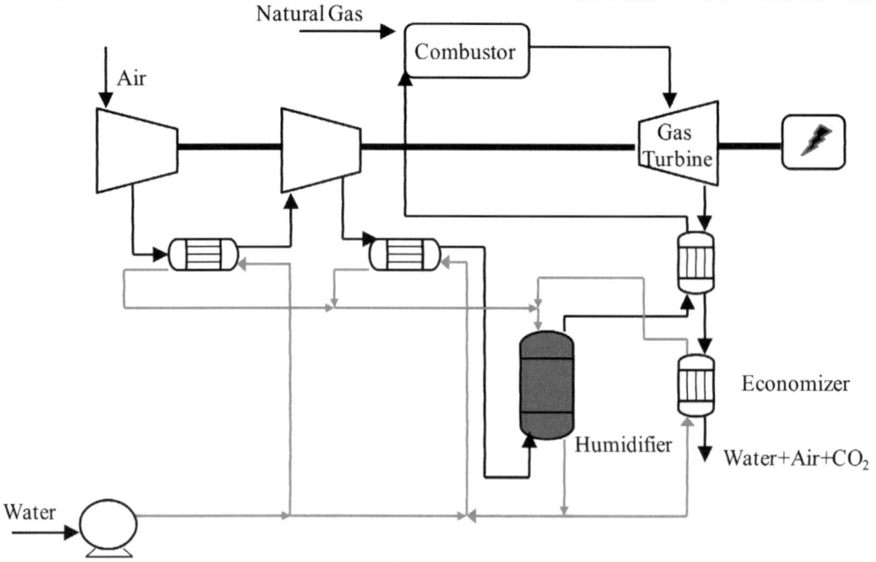

- ***Increase gas turbine output by using humid air as the working fluid. Gas turbine exhaust is used to generate steam which mixes with the compressed air***

14AIChE Annual Meeting, Atlanta, GA

Intercooling, Reheating and Regeneration

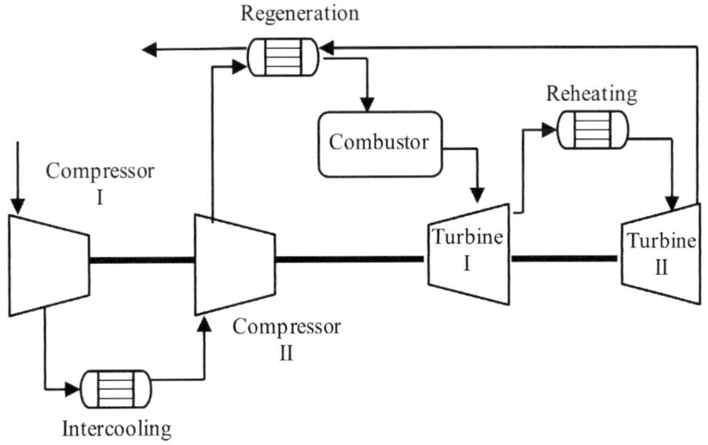

- *Work ratio of a gas-turbine cycle improves as a result of intercooling and reheating*
- *Thermal efficiency is reduced unless they are accompanied by regeneration.*
- *Improvements come at the expense of increased initial and operation costs*

14AIChE
Annual Meeting, Atlanta, GA

2 Simulation Runs

14AIChE
Annual Meeting, Atlanta, GA

Simulations

- 3 CLC systems: Simple CLC, exCLC, CLC3
- 3 Power generation configurations: SC, STIG, HAT
- Two carriers: Fe, Ni

↓

15 configurations, since CLC3 requires multiple oxidation states (only Fe as carrier).

The study also includes 9 operation conditions for each configuration:

- 3 levels of pressure in reactors: 10, 20 y 30 bar
- 3 values of Turbine Inlet Temperature (TIT): 1050, 1200 y 1350 °C
- Main estimations: Cost and net efficiency with respect to fuel Low Heating Value (LHV)

14AIChE Annual Meeting, Atlanta, GA

Simulations

Configuration	CLC	Carrier	Power Cycle
1	Simple CLC	Ni	Steam Cycle
2	Simple CLC	Ni	STIG
3	Simple CLC	Ni	HAT
4	Simple CLC	Fe	Steam Cycle
5	Simple CLC	Fe	STIG
6	Simple CLC	Fe	HAT
7	exCLC	Ni	Steam Cycle
8	exCLC	Ni	STIG
9	exCLC	Ni	HAT
10	exCLC	Fe	Steam Cycle
11	exCLC	Fe	STIG
12	exCLC	Fe	HAT
13	CLC3	Fe	Steam Cycle
14	CLC3	Fe	STIG
15	CLC3	Fe	HAT
16			NGCC

14AIChE Annual Meeting, Atlanta, GA

Models and Modeling Assumptions

Environment: Aspen Plus Process Simulator V7.1

Thermodynamic Models

Peng Robison Equation of State and the SOLID Aspen Thermodynamic Model

Equipment	Aspen Model	Parameters
Pumps	PUMP	Efficiencies: Isentropic 0.70 Mechanical 0.98
Compressors	COMPR	Efficiencies: Isentropic 0.85 Mechanical 0.99
Turbines	COMPR	Efficiencies: Isentropic 0.90 Mechanical 0.98
Heat Exchangers	HEATX	Minimum Approach: 10 °C
Heaters/Coolers	HEATER	Minimum Approach: 10 °C
Reactors	RGIBBS	Chemical Equilibrium
Mixers/Splitters	MIXER/ FSPLIT	
Humidifier	RADFRAC	10 Equilibrium stages

14AIChE
Annual Meeting, Atlanta, GA

Additional Specifications

14AIChE
Annual Meeting, Atlanta, GA

Extended CLC

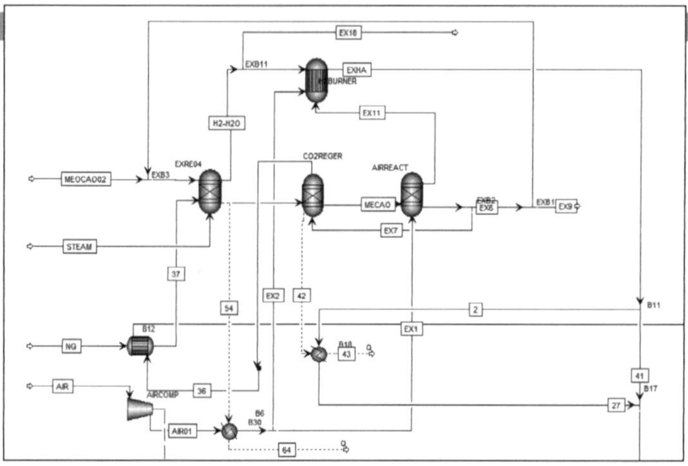

- **Optional: A fourth reactor is an H_2 combustor for the purposes of cost and energy comparisons**
- **Maintaining the calcination reactor at 10 bar and 870 °C, practically pure CO_2 is obtained. Further , the reduction reactor requires 750 °C for H_2 production**
- **Energy obtained from the H_2 combustor is used to achieve those conditions**

14AIChE
Annual Meeting, Atlanta, GA

CLC3

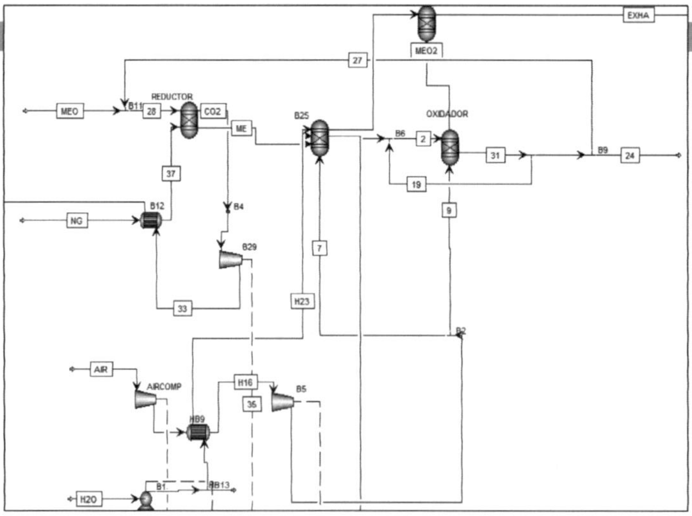

- *Once again, a fourth reactor is an H_2 combustor for the purposes of cost and energy comparisons*
- *A CO_2 compressor is also included*
- *Intercooling between the air compressors generates the steam needed in the oxidation reactor*

14AIChE
Annual Meeting, Atlanta, GA

HRSG

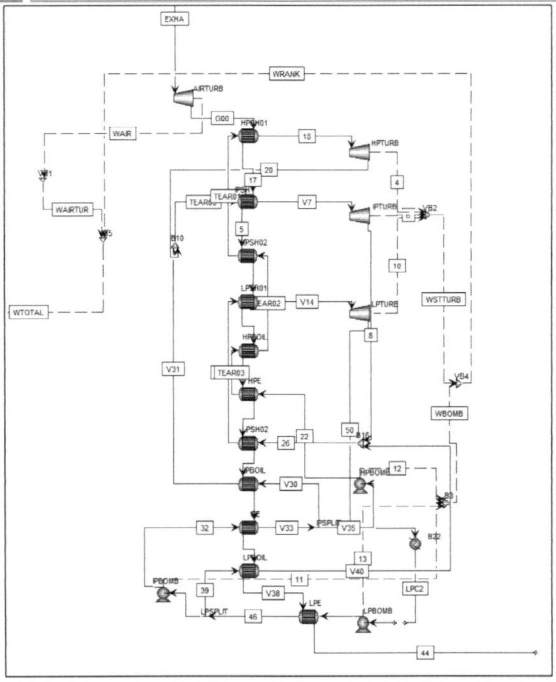

- ***HRSG includes 3 pressure levels for steam generation (1700, 600 and 70 psia)***

14AIChE
Annual Meeting, Atlanta, GA

Additional Specs

- ***Fuel Low Heating Value***

Component	Molar Fraction	Molar Flow (kmol/hr)	Combustion Enthalpy (KJ/mol)	Low Heating Value (MW)
Metane	0.8937	1,622.07	-802.6	-361.63
Etane	0.1011	183.50	-1428.6	-72.82
Propane	0.0052	9.44	-2043.1	-5.36
Total		1,815.00		-439.80

14AIChE
Annual Meeting, Atlanta, GA

Simulations

- Integration of streams and operation conditions is needed between the CLC system and the power generation plant.
- exCLC-HAT

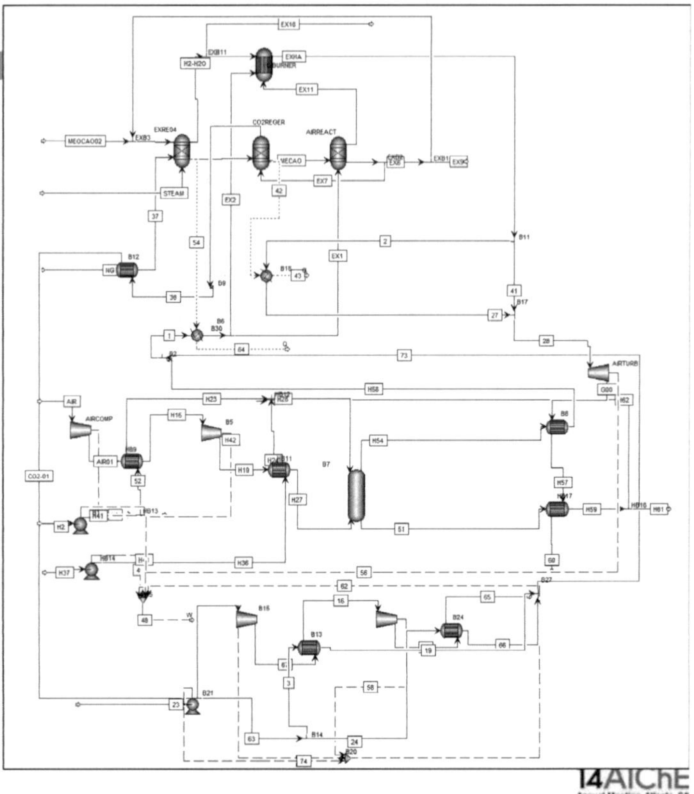

14AIChE Annual Meeting, Atlanta, GA

Additional Specifications

- AspenTech **Design Specs** Tool was used to achieve specific operational conditions. For instance:

  a) Given the fuel stream and the TIT, estimate the air flow required

  b) Estimating the flows among the reactors on the exCLC to achieve adiabatic conditions on the oxidation reactor

- AspenTech **Sensibility Analysis** Tool is used to achieve the highest net LHV plant efficiency (**optimization**)

14AIChE Annual Meeting, Atlanta, GA

Costs

Configuration	Carrier	Composition (% mol)	Price (USD/Kg)
CLC - NI	NiO/Alúmina	0.25/0.75	24,14
CLC- Fe	Fe2O3/Alúmina	0.75/0.25	20,10
exCLC Ni	NiO/CaO/Alúmina	0.28/0.16/0.56	24,57
exCLC Fe	Fe2O3/CaO/Alúmina	0.36/0.10/0.54	20,18
CLC3	Fe2O3/Alúmina	0.60/0.40	20,19

- Aspen Economic Evaluation Tool (Based on a Guthrie type of methodology)
- Cost of utilities uses information from the Mexican Petroleum Industry

14AIChE
Annual Meeting, Atlanta, GA

3 Results

14AIChE
Annual Meeting, Atlanta, GA

Net Efficiencies

- ***AVERAGE Net LHV Plant Efficiency***

Configuration	CLC	Carrier	Power Cycle	%η
1	Simple CLC	Ni	Steam Cycle	45.92
2	Simple CLC	Ni	STIG	47.40
3	**Simple CLC**	**Ni**	**HAT**	**53.21**
4	Simple CLC	Fe	Steam Cycle	47.45
5	Simple CLC	Fe	STIG	48.08
6	**Simple CLC**	**Fe**	**HAT**	**53.80**
7	exCLC	Ni	Steam Cycle	50.29
8	exCLC	Ni	STIG	46.01
9	**exCLC**	**Ni**	**HAT**	**51.28**
10	exCLC	Fe	Steam Cycle	46.54
11	exCLC	Fe	STIG	45.02
12	exCLC	Fe	HAT	45.53
13	CLC3	Fe	Steam Cycle	48.87
14	CLC3	Fe	STIG	46.44
15	**CLC3**	**Fe**	**HAT**	**52.76**
16			NGCC	45.50

14AIChE Annual Meeting, Atlanta, GA

Maximum Net Plant Efficiencies

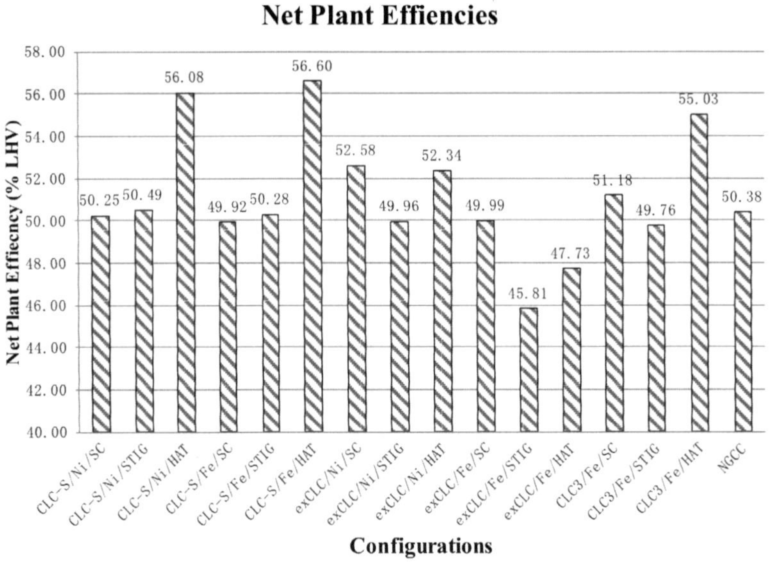

- HAT Power Plant + Nickel provides the best potential for a CLC based configuration. The highest the TIT, the highest the η

14AIChE Annual Meeting, Atlanta, GA

Net Efficiencies

- *HAT Configurations involve higher efficiencies*

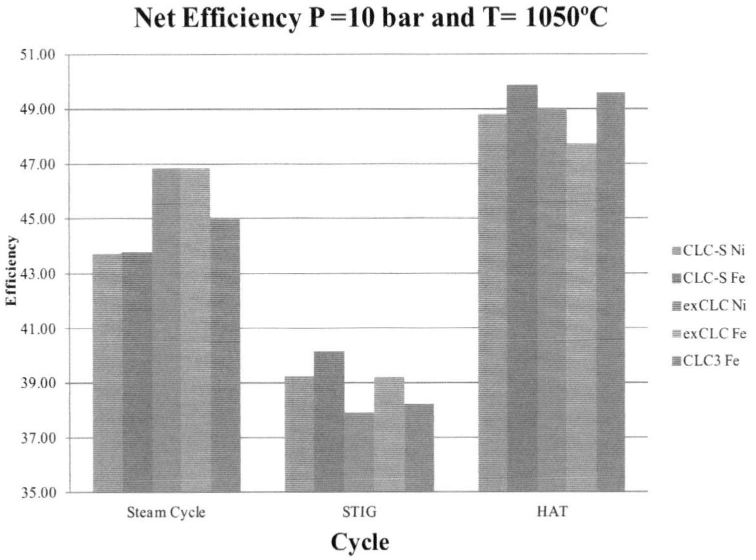

14AIChE Annual Meeting, Atlanta, GA

Effect of TIT and Reactors Pressure

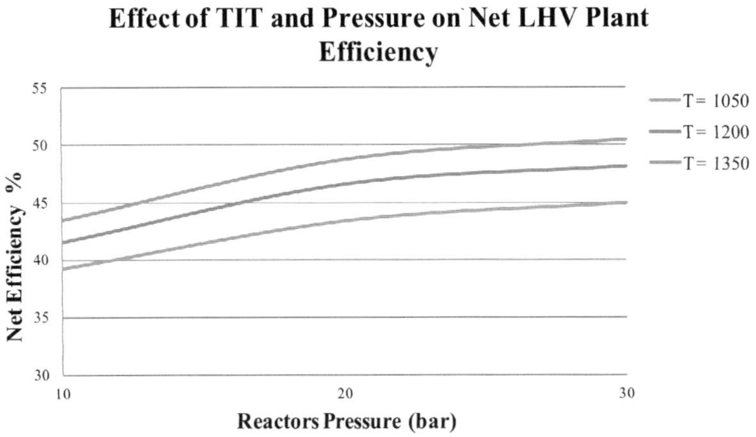

Standard Behavior
CLC+STIG (Ni)

14AIChE Annual Meeting, Atlanta, GA

Effect of TIT and Reactors Pressure

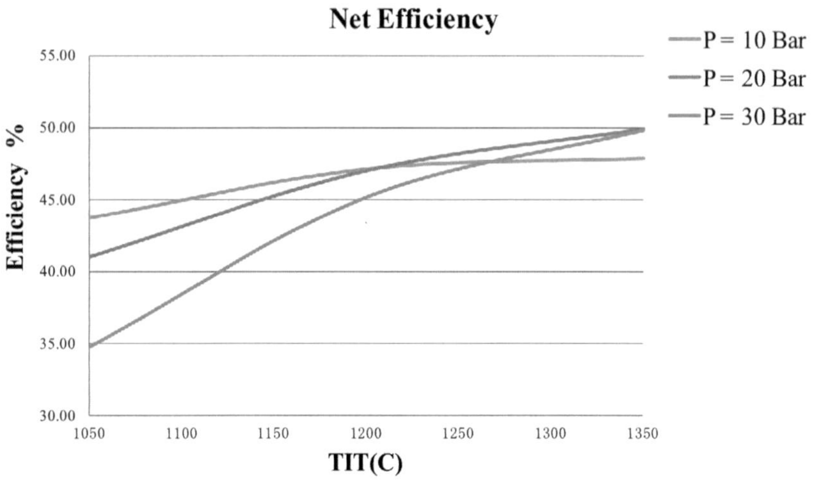

NGCC

Production Cost

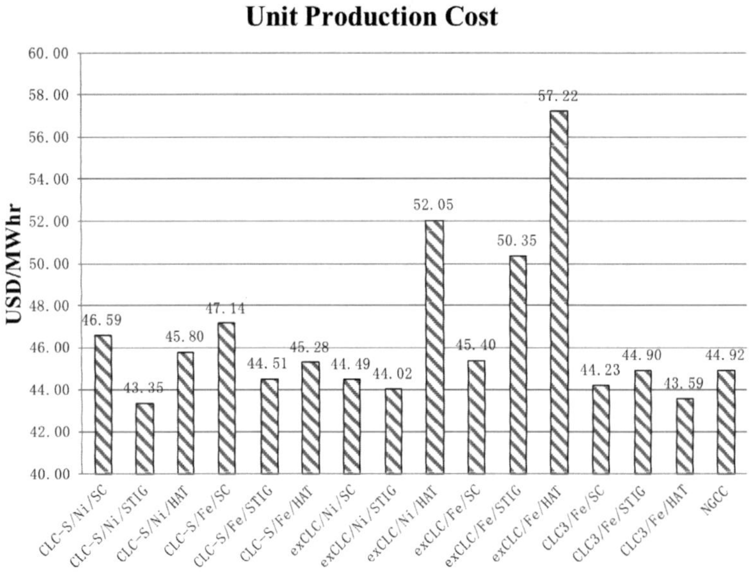

- exCLC configurations involve the highest production cost

4 Discussion and Final Remarks

14AIChE
Annual Meeting, Atlanta, GA

Summary

- Results just confirm the potential of CLC systems for its integration to power generation cycles.
- Simulation also reveals the many operational complications and assumptions that need to be considered, analyzed and validated in practice.
- Under the same basis for comparison, net LHV plant efficiencies obtained in CLC-Power generation plants are competitive to currently used technologies with the advantage of its potential for CO_2 capture

14AIChE
Annual Meeting, Atlanta, GA

Summary

- Unit production costs of energy in CLC based power plants are also comparable to conventional technologies
- From the configurations considered, those involving HAT presented the highest efficiencies
- STIG configurations show the lowest efficiencies
- Configurations involving exCLC show the highest unit production cost of energy
- Configuration involving a Simple CLC with HAT power generation technology is best configuration on this study

14AIChE
Annual Meeting, Atlanta, GA

Special Thanks

- *Urmila Diwekar (UIC)*
- *Moisés Petriz-Prieto (Instituto Tecnologico de Celaya, Mexico)*

14AIChE
Annual Meeting, Atlanta, GA

Thanks !

Questions?

14AIChE
Annual Meeting, Atlanta, GA

AUTHOR INDEX

AUTHOR INDEX

AUTHOR INDEX

AIChE
120 Wall Street, FL 23
New York, NY 10005-4020

CURRAN ASSOCIATES INC.
proceedings.com

ISBN 978-1-5108-1251-2
9 781510 812512